Changing Landscape of Academic Women's Health Care in the United States

INTERNATIONAL LIBRARY OF ETHICS, LAW, AND THE NEW MEDICINE

VOLUME 48

For other titles published in this series, go to
http://www.springer.com/series/6224

William F. Rayburn · Jay Schulkin
Editors

Changing Landscape of Academic Women's Health Care in the United States

Springer

Editors
William F. Rayburn
Department of Obstetrics and Gynecology
University of New Mexico
Copper Rose St. NE., 6112
87111 Albuquerque, New Mexico
USA
wrayburn@salud.unm.edu

Jay Schulkin
Georgetown University Medical School
The American College of Obstetricians
and Gynecologists
409 12th Street SW
20024 Washington, DC
USA
jschulkin@acog.org

ISSN 1567-8008
ISBN 978-94-007-0930-0 e-ISBN 978-94-007-0931-7
DOI 10.1007/978-94-007-0931-7
Springer Dordrecht Heidelberg London New York

Library of Congress Control Number: 2011925158

Printed on acid-free paper

Springer is part of Springer Science+Business Media (www.springer.com)

Foreword

Over the past two decades great advances have been made as women's health issues continue to be recognized as more of a national priority. The result has been an increase in programs and funding examining issues that encompass a woman's unique biology as well as her sociocultural, economic and physical environment (NWHIC, 2000). These factors affect not only the duration but the quality of a woman's life.

Yet, against the background of advances, many impediments to comprehensive healthcare for women persist. Academic medicine includes education and training, research, and clinical care and, as a result, has broad scope and influence on the health system. As we look to the future of significant healthcare reform, we must address the projected demand for healthcare services, the paucity of primary care physicians and sub-specialist in Women's health as neglected components of health professions.

The material in the book attests to both the extraordinary sophistication of the medical system and the challenges we face as we focus on training the next generation of academic clinicians and scientists. More than a century after Flexner wrote his report for the Carnegie Foundation, physicians trained in women's health will practice in a health care delivery system that faces mounting pressures for cost containment, quality improvement and public accountability.

Health care providers must raise awareness of the importance of women's healthcare. The US Secretary of Health and Human Service, through the Council on Graduate Medical Education (COGME) as its advisory body, recommends physicians develop a broad understanding of health conditions specific to Women. In this book the authors have synthesized a wealth of information into an accessible format, and have a host of practical recommendations that address the challenges of this changing landscape.

The authors have succeeded admirably and I am grateful to them for all of their efforts and I highly recommend this book. While this book cannot address the entire range of complex issues in the arena of women's health, the authors have put an impressive group of thought leaders to discuss key trends in the U.S. health care delivery system that effect how women's healthcare is delivered. It is my hope that barriers will be further diminished as health care professionals, medical administrators, medical directors, medical school faculty utilize the timely information within this book.

New York Mary E. D'Alton

Preface

Since 2005, a dozen of the United States and more than 15 medical specialties have reported a physician shortage or anticipate one in the next few years. This anticipated shortage and a worsening of physician distribution are compounded by a projected increased demand for health care services. Health care of women of all ages is particularly vulnerable. The obstetrician gynecologist workforce is aging and is among the least satisfied medical specialists. Primary care physicians such as family physicians and general internists will be asked to care for more women who are elderly. Furthermore, fellowship training in women's health care in internal medicine and in maternal child health in family and community medicine involves only a small portion of general internists and family physicians.

As we cope with significant health care reform in the United States, academic departments of obstetrics and gynecology, family and community medicine, and internal medicine have opportunities to create a unified women's health curriculum for undergraduate students, share preventive health and well-woman expertise in training programs, provide improved coordination of care, and instill concepts of lifelong learning to our graduates.

To address this changing landscape, this text contains chapters that focus on strategic planning on behalf of academic faculty who will train the anticipated additional load of students and those residents and fellows with particular interest in women's health care. We begin with a look back at the Flexner report in which medical schools a century ago did not base training on sufficiently high standards of science. Recommendations presented here from authors with distinguished leadership skills indicate a consensus, but not unanimity. We summarize in the final chapter our collective expertise and offer ways to implement recommendations to better prepare for tomorrow's needs in academic women's health care.

This unique book provides timely information for health policy professionals, health care professionals, medical administrators, and medical school faculty. We appreciate the forward thinking brought forth by this distinguished group of academic leaders. Their presentation of current information and projections of changes will serve as a foundation for continuing education and further investigation during this time of health care reform.

Albuquerque, New Mexico
Washington, District of Columbia

William F. Rayburn
Jay Schulkin

Contents

Contributors

Mary C. Blehar Consultant to Office of Research on Women's Health, formely Chief of Women's Programs, National Institute of Mental Health, National Institutes of Health, Potomac, MD, USA, blehar.mary@gmail.com

Breda Bova 1 University of New Mexico, Albuquerque, NM, USA, bova@unm.edu

Kathleen C. Bowling Department of Obstetrics and Gynecology, Warren Alpert School of Medicine of Brown University, Providence, RI, USA, KBowling@WIHRI.org

Joanna M. Cain Department of Obstetrics and Gynecology, Warren Alpert School of Medicine of Brown University, Providence, RI, USA, joanna.cain3@gmail.com

Eric J. Hodgson Division of Maternal-Fetal Medicine, Department of Obstetrics, Gynecology and Reproductive Sciences, Yale University School of Medicine, New Haven, CT, USA, eric.hodgson@yale.edu

Colleen M. Horan Department of Obstetrics and Gynecology, Jefferson Medical College, Thomas Jefferson University, Philadelphia, PA, USA, Colleen.horan@jefferson.edu

Julia V. Johnson Department of Obstetrics and Gynecology, University of Massachusetts Memorial, Worcester, MA, USA, Julia.johnson@umassmemorial.org

Charles J. Lockwood Department of Obstetrics, Gynecology and Reproductive Sciences, Yale University School of Medicine, New Haven, CT, USA; Obstetrics & Gynecology, Yale-New Haven Hospital, New Haven, CT, USA, charles.lockwood@yale.edu

Anne Drapkin Lyerly Department of Social Medicine and Center for Bioethics, University of North Carolina School of Medicine, Chapel Hill, NC, USA, alyerly@email.unc.edu

Kristen A. Matteson Division of Research, Department of Obstetrics and Gynecology, Women and Infants Hospital, The Warren Alpert School of Medicine of Brown University, Providence, RI, USA, kmatteson@WIHRI.org

Sharon Phelan Department of Obstetrics and Gynecology, University of New Mexico, Albuquerque, NM, stphelan@salud.unm.edu

Vivian W. Pinn Office of Research on Women's Health, Department of Health and Human Services, National Institutes of Health, Bethesda, MD, USA, vivian.pinn@nih.gov

William F. Rayburn Department of Obstetrics and Gynecology, University of New Mexico, Albuquerque, NM, USA, wrayburn@salud.unm.edu

Jay Schulkin Department of Research, The American College of Obstetricians and Gynecologists, Georgetown University Medical School, Washington, DC, USA; Department of Physiology, Biophysics, and Neurosciences, Georgetown University Medical School, Washington, DC, USA, jschulkin@acog.org

Erin E. Tracy Vincent Obstetrics and Gynecology, Massachusetts General Hospital, Boston, MA, USA, EETracy@partners.org

Louis Weinstein Department of Obstetrics and Gynecology, Jefferson Medical College, Thomas Jefferson University, Philadelphia, PA, USA, Louis.weinstein@jefferson.edu

Chapter 1
The Flexner Report and Evidence-Based Medicine Over the Past 100 Years

Jay Schulkin and Kristen A. Matteson

Abstract The Flexner Report began a process in North America linking science with physician education and practice. While the report emphasized the importance of standards of clinical training and care and the role of the laboratory in physician education, this chapter deals with the explosion of knowledge that pervades decision-making in all medical disciplines, especially those bridging women's health care.

Keywords Flexner · Evidence-based medicine · Medical decision making · Medical culture

1.1 Introduction

The management of discomfort is an ancient practice. All human beings seek to ameliorate pain, prevent disease, and treat disease should it occur. Human evolution and adaptation to address this very basic need is at the very foundation of medical decision making [1].

Medical decision making has evolved and developed over the course of history. Instrumental in this process were the Flexner brothers, who fueled change in the medical education system in the early 20th century, and were intrigued by the scientific education provided by German universities and the concept of integrating research and medical education. The legacy of the Flexner brothers and the Flexner Report on medical education in the 20th century was the incorporation of research and scientific evidence into the clinical encounter; because of changes promoted in the Flexner Report, clinicians and patients now discuss the evidence within a

J. Schulkin (✉)
Department of Research, The American College of Obstetricians and Gynecologists, Washington, DC, USA; Department of Physiology, Biophysics, and Neurosciences, Georgetown University Medical School, Washington, DC, USA
e-mail: jschulkin@acog.org

W.F. Rayburn, J. Schulkin (eds.), *Changing Landscape of Academic Women's Health Care in the United States*, International Library of Ethics, Law, and the New Medicine 48, DOI 10.1007/978-94-007-0931-7_1, © Springer Science+Business Media B.V. 2011 medical culture that promotes evidence-based clinical practice and patient participation in decision-making. Obstetricians and gynecologists, skeptical of research and scientific evidence in the late 19th century, have embraced incorporating sound scientific evidence into clinical practice over the past several decades.

This chapter begins with a brief depiction of the "science" behind human problem solving abilities, which form the basis for medical inquiry. Next, we discuss how the German medical education system and the Flexner brothers influenced the development of what is considered modern medical inquiry and standardized medical education. We then look at how medical inquiry and the field of obstetrics and gynecology co-evolved during the 20th century to address the changing landscape of medicine, including modern medical decision-making, evidence based medicine, the culture of shared decision-making between physicians and patients, continuing medical education, and job satisfaction.

1.2 Background: The Development of the Human Propensity for Problem Solving

Human beings have a cognitive predilection to make associations between objects and events through self-corrective inquiry; that is, we look to determine relationships between events, to form hypotheses, and to draw conclusions.

The construction of objects, a fundamental adaptation of our species, expanded from our first small tool use, likely barely modified natural objects like sharpened sticks and bashed pebbles, to more elaborate tool design and utilization. Cephalic expansion, the evolution of the brain, expanded the ability of humans to experience the world and facilitated technological creations. The development and use of magnifying devices, for instance, helped humans move from a situation where nature was "managed" to a situation where nature was "understood" and "manipulated". The evolution of tool making and tool use were both critical for this development [2].

Tool use is an expression of our evolved brain and its underlying cognitive systems. There are several examples of this throughout our human history: Galileo made his discoveries of the starry heavens because of the telescopic lens that expanded sight, the development of the microscopic lens allowed van Leeuwenhoek to see bacteria and yeast, Laeneec developed the stethoscope to facilitate auscultation of the heart and lungs, and Chamberlen invented forceps to assist with delivering babies from mothers with obstructed labor (Fig. 1.1) [3]. Expanded tool use, along with the development of the experimental method, set the stage for the evolution of the medical decision-making process. Though the stage was set, centuries would pass before the Flexner brothers detailed medical decision-making and the integration of experimentation within a model of medical education, concepts that underlie the Flexner Report [4].

Tool development and use represents one aspect of human cognitive development. A major cognitive achievement of humans is the desire to provide relief from distress. Most animals, in comparison to humans, have a limited range of

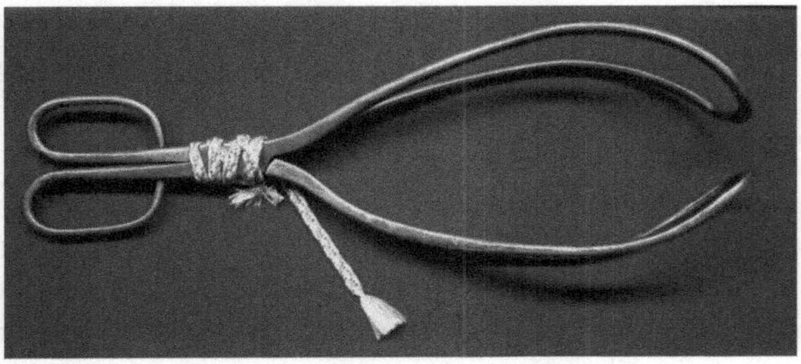

Fig. 1.1 Chamberlen forceps

responses to the pain of others [5]. Evidence of what we now consider medical decision-making has been consistently documented in historical texts in the form of algorithms to temper the distress of others. Although determining causal relationships pertaining to health and bodily function is a concrete task, general decision making involves an abstract and diverse set of cognitive resources.[1]

Humans are, by predilection, taxonomic animals; the categorizing function of cephalic adaptation forges coherence about objects, kinds of events, and entities. Aristotle was a dominant early expositor of this predilection [6]. The taxonomic tradition handed down from Aristotle to Galen, through Ray, Linnaeus, and Buffon, along with the experimental approach and the invention of new technologies, paved our growing understanding of natural diseases.[2] It is no surprise that early on in our cultural history cataloging would be manifest in maintaining health and warding off disease.

The field of medicine involves many different layers and levels of problem-solving and different kinds of approaches ranging from bedside to laboratory. Specific types of problem-solving and approaches to obtaining medical information are presented in Table 1.1. This range of approaches within medical problem solving developed into scientific and medical inquiry and set the stage for structured modern medical education, detailed in the next sections of this chapter [7].

Table 1.1 Characteristics

Kinds of Medicine	Object of Inquiry	Form and Site of Education	Goal
Bedside	Whole Patient	Apprenticeship	Therapy
Library	Text	Scholastic, Linguistic, University	Preservation, recovery, commentary
Hospital	Patient, organ	Hospital	Diagnosis
Social	Population, statistic	Community	Prevent
Laboratory	Animal Model Cells	Laboratory	Understand

1.3 Gaps in American Medical Education in the 19th Century: Looking to Europe and Setting the Stage for Medical Education Reform

While there has always been resistance to change in medical training and education, physicians and medical researchers have continuously questioned the methods used to train medical students, and eventually came to accept the need for a standardized system of medical education. The culture of medicine evolved over these last 100 years to become accountable to its patients and students [8]. The need to reform medical education in the United States became apparent during the Civil War (1881–1865), when more soldiers died from infections than from actual fighting, and when physicians rarely used available medical instruments (such as the stethoscope) but often employed traditional medical therapies which had been proven harmful [9]. Skepticism about new medical evidence and technologies was not unique to the battlefield. During this same time period, puerperal fever was a major cause of maternal morbidity and mortality within hospitals [10]. Despite this, when Oliver Wendell Holmes (1809–1894)and Ignaz Semmelweiss (1819–1865) scientifically proposed that puerperal fever was contagious and carried from patient to patient by physicians they were disbelieved and opposed.

American medicine started as a trade practice divorced from the rigors of science. Most medical schools in the United States operated without any concrete affiliations with hospitals or universities, and began enrolling students in the early 1800s as a supplement to the traditional apprenticeship system. Most of these medical schools were independent businesses run by the professors of the school, and large enrollments were necessary to ensure salaries and profits to the professors. As a result, admission standards were essentially non-existent and an incredibly small minority of students had a high school education, let alone a college degree.

Though there were exceptions, the overwhelming majority of physicians attended schools that were not providing adequate training. Although this system allowed the United States to train many physicians and meet the growing needs of its expanding population, it did not ensure adequate care, especially when compared to medical education in Europe. It was not until University of Pennsylvania opened America's first university-affiliated medical school (1765), followed by Harvard (1782), that trends in medical education in the United States began to shift. At these institutions, medicine evolved as a profession with direct links to science.

Beginning in the early 1800s, physicians who were interested (and had the financial means) to pursue post-graduate training studied in Europe. In the first half of the century, Paris was considered the best place for these students to train, and they would learn the French methods of medical inquiry. However, French practice before the dominance of Claude Bernard tended to rely solely on astute observation [11]. In the second half of the 19th century, Germany emerged as a scientific and medical powerhouse, and its success is mostly attributed to its use of experimental methods and its integration of the basic sciences (such as biochemistry and microbiology) with clinical medicine and medical knowledge. Most students who returned to the U.S. from their training in Europe could not find faculty medical positions that allowed them to integrate their interests in research and medical discovery with teaching and clinical responsibilities. However, some of the talented students returning from Europe became medical school faculty at prominent institutions, and had immeasurable influence on the shape of medical education at the beginning of the 20th century and on the modern medical education we recognize today. The tradition of using empirical science to guide medical practice eventually arrived in the US; The Johns Hopkins University School of Medicine opened in 1893, and carefully incorporated the German practice of well-respected scientific inquiry into its medical school curriculum.

The Johns Hopkins University School of Medicine was the first real medical research institution in the U.S. William Osler (1849–1919), a pathologist trained in Europe and the first chief of staff at Johns Hopkins Medical School, is sometimes called the father of modern medicine [12]. He wrote *The Principles and Practice of Medicine: Designed for the Use of Practitioners and Students of Medicine*, published in 1892, which served as one of the most significant and widely used medical textbooks in the United States for more than 40 years. Besides Osler, Johns Hopkins boasted three other prominent physicians who would have major influences upon the development of modern academic medicine: William H. Welch, William S. Halsted, and Howard Atwood Kelly.

William H. Welch (1850–1934), who trained in Germany and was determined to be a full-time researcher and teacher, was a pathologist. He was the first dean of The Johns Hopkins University School of Medicine, and the founding president of the Board of Scientific Directors at the Rockefeller Institute for Medical Research. William S. Halsted (1852–1922), a good friend of Welch and the first chief of surgery at Hopkins, is credited with starting the first formal surgical residency training program in the United States; he introduced numerous innovations in surgical procedures including techniques for inguinal hernia repair and mastectomy. Also renowned in his field, Howard Atwood Kelly (1858–1943), perhaps the first respected gynecologist in the U.S. had great interest in and influence on the understanding of female reproductive anatomy [13]. The system of medical education embraced by Johns Hopkins was the foundation upon which the Flexner Report was written.

Medical education, and therefore education in women's health and obstetrics and gynecology, must evolve with changing scientific knowledge. However, for medical education to truly evolve, physicians, administrators, researchers, students, and trainees must be willing to critically examine not only what knowledge is being shared, but the processes by which it is being taught. Two important individuals during this time period who questioned how medical knowledge was being acquired and shared with trainees were Simon and Abraham Flexner.

1.4 Reforming Medical Education to Highlight the Importance of Evidence and Problem-Solving: Abraham Flexner, the Flexner Report, and Medical Education Over the Past 100 Years

Despite their exceptional influence on modern medical education, the Flexners and their work are not very well known today, even among academics. Simon Flexner (1863–1946) was a physician, scientist, administrator, and the first director of the Rockefeller Institute for Medical Research (where he met William Welch of Johns Hopkins). Simon Flexner became passionate about science because of an illness he suffered in childhood, and he taught himself to use a microscope in his brother's pharmacy. Because of these formative personal experiences, he developed and professed the belief that science had no boundaries and that every individual should have access to it and be able to understand it. His research focused on pathology, bacteriology and virology, and he played an essential role in developing treatments for meningitis, which was epidemic in the early 1900s [14]. During this time, poliomyelitis also became epidemic and Flexner and his research team proved its viral origins and mode of transmission, laying the groundwork for the development of the polio vaccine in the 1950s, a true success story of modern medicine and scientific inquiry.[3] Flexner's approach to inquiry, taking his expertise in meningitis and applying it to studies on poliomyelitis, played a large role in this public health breakthrough.

But it was Abraham Flexner (1866–1959), who had been an undergraduate at Johns Hopkins and then went back to teach high school in his home town, who would be given the task of investigating medical schools in the United States and providing a detailed description of the state of medical education. Abraham had founded several experimental and highly regarded private schools, and he also served on the Rockefeller Foundation's General Education Board. In 1908 his first published work, *The American College,* attracted the attention of the Carnegie Foundation, which appointed him to lead a study of professional education in the U.S. Abraham visited 131 American and 24 Canadian medical schools in 1910, and organized an official report, making recommendations for change in medical education.[4] His critical journalistic approach looked at both the social and scientific aspects of medical education at the time [15].

Abraham Flexner found that many schools were ill-equipped to prepare students to become physicians. Standards for education and admission were wildly inconsistent and only few schools were truly outstanding. Few medical schools even required a college degree as a prerequisite of entrance. Flexner suggested the closing of many schools and the restructuring of most of the rest. His main focus was on the structure of medical education and he based many of his recommendations on psychological theory and the structure of medical education in Germany and at Johns Hopkins. Flexner emphasized the need for a solid foundation in the basic sciences, combined with training in research techniques, and a clinical phase of education in order to develop students' problem solving skills as they pertained to dealing with actual

Table 1.2 Flexner's impact on U.S. and Canadian schools [16]

Did not close or merge as a result of the Report	**97**	**58%**
Closed or merged because of the Report	**12**	**7%**
Closed	9	5%
Merged	3	2%
Closed or merged for unclear reasons	**59**	**35%**
Closed or merged in the two decades after the Report	26	15%
Closed or merged more than two decades after the Report	3	2%
Unknown	30	18%
Total	**168**	**100%**

None of the eight Canadian schools closed or merged as a result of the Report

patients. The Flexner report advocated the idea of a uniform, integrated medical school education across the country.

The closing or merging of 12 medical schools can be directly attributed to the Report, and additional 26 schools closed for unclear reasons in the 20 years following its publication (Table 1.2) [16].

At the time of the Flexner report, only 921 women were enrolled in medical schools, a decline from the previous year when enrollment was over 1400 [17]. This has dramatically changed in the 100 years since his report [18]. Although there were few female physicians or women in medical schools when Abraham Flexner was constructing his report in the early 1900s, his report likely played an unfortunate role in actually decreasing the already miniscule numbers of women and African Americans pursuing careers as physicians as many of the schools that were closed as a result of the report were the schools most likely to offer the opportunity of medical training to women and minorities [19].

Flexner's basic premises laid the groundwork for modern medical inquiry and medical education. He emphasized that physicians and physicians-in-training needed to be comfortable with truly inquisitive processes and that these processes needed to span everything from laboratories to the bedside [20]. Since Flexner wrote his report, of course, we have witnessed a profound expansion of information and knowledge in basic sciences, clinical sciences, and medicine. Now, we commonly accept that learning is a lifelong process [21].[5] In order to effectively teach future physicians and physicians in practice, one must consider the goals of the learner, different ways of teaching skills, and faculty development needs. Each of these concepts were strongly supported in Flexner's report and placed in the context of the broad and evolving material that obstetrician gynecologists and women's health care providers need to be aware of in their practice [22].[6]

The medical schools that were not closed after the report attempted to incorporate Flexner's suggestions and devised curricula that were divided into pre-clinical and clinical teaching within departments. This division within academic medical departments may in fact have hindered the realization of Flexner's ultimate objective for medical education: students' development of formal integrated analytic reasoning. In the first half of the 20th century, division and competition for resources between pre-clinical and clinical years, between clinical departments (such as surgery and internal medicine) and within departments (for instance endocrinology and gastroenterology) resulted in compartmentalized medical education.[7]

The incorporation of structured clinical rotations and "hands-on" learning for medical students addressed Flexner's strong recommendation for a clinical phase of medical training, however the fragmentation of medical school curricula failed to optimize the development of analytic thinking. Compartmentalized education was identified as a problem as early as the 1920s and began to be addressed in earnest in the 1950s and 1960s when schools began revising their curricula so that knowledge absorbed in the pre-clinical and clinical years and from various clinical disciplines were integrated.

Medical education must be constantly questioned, assessed, and reconfigured in order to meet the demands of society, patients, students, and ourselves. Knowledge about medicine and the context within which medical care is delivered is ever-changing. Flexibility to change the structure of medical education and medical practice is necessary to ensure that we maintain a pre-eminent medical system by optimizing physician training and retaining highly qualified physicians in their medical fields.

The importance of constantly evaluating and guiding the evolution of medical education is emphasized in The New England Journal of Medicine's medical education series from 2006. In "American Medical Education 100 years after the Flexner Report", Cooke et al stated the overarching goal of medical education as follows: "the final test of [students'] efforts, however, will be not what they know but what they do" [23]. For students and residents to develop into physicians who provide comprehensive care for their patients, obtaining a sound knowledge base is not enough. Trainees need to learn both how to translate their knowledge into methodical clinical care and how to deliver this care professionally and with compassion. Though these skills can be discussed in a pre-clinical or classroom setting, they require guided and supervised interactions with real patients to be nurtured and developed. Facilitating meaningful interactions between students and patients, an instrumental element in formal medical education, is difficult with increasing requirements for supervision, patient safety, and medical record documentation.

Medical education represents life-long learning beyond medical school and residency training. Continuing medical education programs, in order to be successful need to emphasize diverse forms of cooperative learning and interactions where integrative medicine and research are at the forefront. This concept is incorporated in recommendations from the Carnegie Foundation for higher education in which there is envisioned a medical education system that maximizes flexibility in the process of achieving standardized outcomes, creates opportunities for integrative and collaborative learning, inculcates habits of inquiry and improvement, provides a supportive learning environment for the professional formation of students and residents, while at the same time it advances the health of patients and patient populations [24].[8]

1.5 Medical Decision Making and Evidence-Based Medicine (EBM): Flexner's Ideals in Practice

Medicine, of course, has changed dramatically since Flexner wrote his report, in part because of the changes in medical education that the report ignited, but also because of the surge of scientific knowledge in the past 100 years. Although "evidence-based medicine," as we now know it, was formally described in the 1990s as a new way to teach medical thinking, Flexner set the stage for evidence-based medicine and modern medical decision making by sparking the incorporation of original investigation and research into medical education.

This change in not only medical education, but medical culture and medical thinking, became increasingly important with the massive influx of new information, innovation, and scientific discoveries in the medical field in the past century. With this rapidly expanding field of information, it is paramount for physicians practicing in the 21st century to have the ability to critically review available medical evidence, recognize its limitations, and reach an informed decision. In this context, medical decision-making goes hand in hand with evidence-based medicine.

Evidence-based medicine (EBM) arose from epidemiology, which deals with the logic of research design, the clarification of studies, and the broad assessment of data and results. In a real sense, the term "evidence-based" signifies scientific legitimization and involves experts grading information, making it available, and determining whether or not it meets set criteria for quality and importance. EBM provides a scientific foundation for medical decision-making.

EBM is part of the larger culture of evaluating the glut of information and scientifically rigorous research that is available and turning that information into informed choice [25].[9] Perhaps the most widely recognized proponent of EBM was Archie Cochrane, after whom the Cochrane Centre and later the Cochrane Collaboration were named. Cochrane published "Effectiveness and Efficiency" in 1972, six decades after Flexner's report, and made many of the same recommendations. He advocated that health services be evaluated based on scientific evidence rather than tradition or accepted clinical practice.

Ian Chalmers, an obstetrician who worked with and was influenced by Archie Cochrane, was concerned about how quickly technologies were changing in perinatal medicine and how clinical practice was not supported by scientific evidence [26]. In response, he worked to develop a register of perinatal randomized clinical trials and published "A Classified Bibliography of Controlled Trials in Perinatal Medicine 1940–1984" in 1985 [27]. Also identifying a need for systematic scientific reviews in the area of maternal-child health, Chalmers assembled a team to perform systematic reviews to guide the delivery of health care to pregnant women. Since these efforts of Cochrane and Chalmers, teams of experts, such as the Cochrane Collaboration, have worked to make the expanse of medical information manageable for clinicians by evaluating information, deciding on the quality of studies, and determining the strength of the evidence. The founders of medical epidemiology and EBM sought to find rigorous scientific grounding for medical decision making by having the decisions strictly follow the science [28].

Evidence-based material provides physicians with the ability to consult evidence-based analyses of findings from randomized control trials. One well-known version of EBM is the Cochrane library, which evolved in the United Kingdom to provide health care decisions based on sound logical footing and good science [29]. The Cochrane library relies on principles including collaboration, reduction of bias, access to up-to-date information, inclusiveness and goodwill of those involved [30]. Within a relatively short period of time several Cochrane centers developed both within and outside of Great Britain [31].

One primary factor in EBM is the importance of understanding degrees [32]. The primary arguments for developing evidence-based guidelines are to reduce potential for misleading biases, to make the evidence more transparent, and to place decision-making in the context of self-corrective processes. In one of his essays written in the 1970s, Cochrane stated that obstetricians, who made the worst use of clinical trials of any specialists, were the specialists least likely to practice EBM.[10] Physician practices and obstetrician-gynecologists have adapted to embrace EBM over the past few decades [33]. One recent study found that physicians generally rated evidence-based material as very helpful to their practices [34].

Wide variation and inconsistency still exists in the implementation of practice guidelines and in the effectiveness of guidelines in influencing evidence-based decision-making [35]. At least in populations of obstetricians and gynecologists in the United States, evidence-based practice guidelines are perceived as important[11] but keeping track of actual physician implementation of them is not easy. In some contexts obstetricians-gynecologists who read evidence-based documents tend to perform better on self-reported knowledge and practice questions. Surveys of the over 50,000 American College of Obstetricians and Gynecologists (ACOG) fellows consistently suggest that evidence-based materials (such as the ACOG guidelines) are a vital tool in influencing practice and sustaining knowledge. For example, one study found that physicians who were more familiar with the cystic fibrosis (CF) screening document produced by ACOG and the American College of Medical Genetics were more likely to answer knowledge questions about CF correctly (Fig. 1.2). Additionally, it has been shown in other obstetric areas that physicians who were familiar with practice guidelines were more knowledgeable. These findings reflect Flexner's recommendations in action; clinicians need to continue learning and increasing their knowledge base in order to improve the care they deliver to patients.

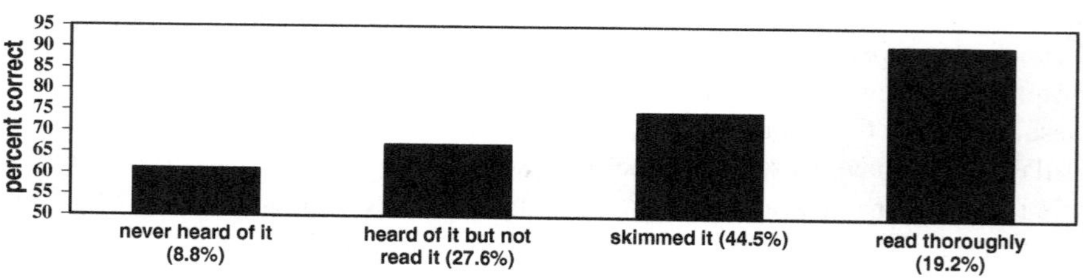

Fig. 1.2 Proportion of physicians answering CF knowledge questions correctly, divided by how intently they read educational materials

However, when fellows were asked how familiar they were with the CF screening document, only 19.2% of physicians said they had read it thoroughly; 44.5% had skimmed it, 27.6% had heard of it but not read it, and 8.8% said they had never heard of it. (see Fig. 1.2) Nevertheless, almost three quarters (72.7%) of doctors said their practice patterns regarding CF carrier screening and prenatal testing had changed since ACOG issued its guidelines.

EBM is an important intellectual development linking research to clinical practice. However, evidence-based reviews from experts are only one tool and must be viewed critically and used properly in order to be useful [36]. Processes for developing guidelines, including peer review of the reviewer, increased awareness of biases, and consideration of the rules for grading the evidence (including the logic of considering the randomized controlled trial as the most highly valued form of medical research) should be evaluated and scrutinized to optimize the quality of evidence and guidelines presented [37].

1.6 Decision Science and Medical Decision Making

The vast literature on decision sciences covers how personal perspectives and biases of the individual making the decision influence the process [38]. One key aspect of medical decision making is accounting for and coping with uncertainty in a rational way [39].[12] Contrasting EBM and decision science, EBM rests on collective choice (asking physicians to accept their colleagues' evidence-based judgments) and decision science focuses on recognizing and accounting for possible bias in decision-making.

The origins of decision science are bound to statistical analysis, defined as the determination of the utility of a decision in the likelihood of considering the outcomes [40].[13] Regardless of whether one is a physician, patient, bus driver, or rocket scientist, individuals are vulnerable to many kinds of errors in reasoning and need to work to self-correct for them.[14] Statistical literacy did not exist in Flexner's depiction of medical education, but it has come to lie at the heart of medical reasoning.

The decision science literature is rich in detail, depicting the diverse ways in which we overvalue or undervalue variables (e.g., whether hormone therapy will result in breast cancer) depending upon our orientation to events and frame of reference [41].[15] This also holds for our memory of events, by which we associate rewards or hedonic value, overvaluing whether, for example, we would be happier living in the state of Washington versus somewhere in the south-west, regardless of whether we actually live in either place [42].

Arthur Elstein, a primary editor and founder of the journal *Medical Decision Making* some 30 years ago, summarized the differences between EBM and decision science, both of which work to enhance medical decision-making (Table 1.3). Both disciplines are vital for a modern Flexnerian approach to continued learning.[16]

Table 1.3 Differences between evidence-based knowledge and medical decision making

Issue	Evidence-based knowledge	Medical decision making
What are the core problems?	Insufficient knowledge	Combining data and values consistently
	Keeping up with the literature	Judgment limitations
Solution	Critical appraisal	Decision tree
	Answerable clinical question	Sensitivity analysis
Formal models	Unimportant	Essential
Decision processing costs	Low	High
Utility assessment	Largely ignored	Core research issue
Cost-effectiveness analysis	Largely neglected	Major application
Decision psychology	Largely ignored	Core research issue

1.7 Shared Decision Making: Patient Rights and Safety

The modern and welcomed shift in thinking within medical practice moves away from the culture and practice of medicine in which the Flexner Report was couched and away from the narrow authority-driven notion that the physician knows what is right and makes medical choices for the patient. Shared decision-making, as it is currently used in medical practice, requires that the patient takes responsibility for decisions that impact her health and that a physician-patient partnership emerges as a normative goal within clinical care [43].

Even as medical marvels have exponentially increased and the complexity of medical knowledge has grown, our culture has valorized respect for and promotion of patient participation and patient responsibility [44]. Just as medical education and medical knowledge has evolved, the "clinical encounter" has evolved as well. With shared decision making, physicians aim to thoroughly inform a patient before the patient chooses her course of treatment. Physicians facilitate this endeavor by trying to make the information as transparent as possible [45]. This process of sharing information is a goal of the culturally reinforced covenant or contract between the physician and the patient, between the woman and the obstetrician-gynecologist [46].

Shared medical decision-making has evolved, placing the physician's evidence-based decision-making and counseling within the context of respect for and promotion of patient participation and patient responsibility for their own medical care.[17] Current patient-physician interactions rely upon physicians interpreting available evidence, providing the evidence necessary for decision-making to the patient, and the patient making an informed decision. For this process to take place, both the clinician and the patient must both have sufficient knowledge on the medical condition and treatment options. Shared decision-making process benefits patients and truly respects their autonomy.[18] An ACOG Committee Opinion on Patient Testing emphasizes the essentiality of the patient in decision making with the following well-written summary:

> The physician and the patient have a shared responsibility. The quality of medical care improves when there is clear communication and mutual understanding between physician and patient. It is the responsibility of the obstetrician-gynecologist to communicate effectively and to develop skills that promote a patient-physician relationship that is characterized by trust and honesty. Similarly, it is the responsibility of the patient to provide accurate information about her lifestyle, health habits, sexual practices, and religious and cultural beliefs when these factors may affect clinical judgement [47].

In order to provide evidence to patients in a transparent fashion, however, physicians must consider both what is known and what is unknown, the biases within the body of literature itself, the personal perspective of the physician, and possible conflicts of interest [48]. To genuinely share the decision-making process, physicians must look beyond their biases and move beyond just the medical evidence to (1) understand the patient's experience and expectations; (2) build trust and participation; (3) provide evidence-based material with a discussion of uncertainties; (4) expose biases in the data and orientation; and (5) continuously check for understanding on both sides [49].[19]

In the end, the physician has a moral responsibility to respect the individual and her right to self-determination, her right to make her own informed medical decision [50].[20] Shared medical decision making embraces a modernized Flexnerian vision of embracing scientific evidence and incorporating it into clinical care and physician/patient communication; it reinforces a partnership between physicians and patients with the goal of enhancing medical care.

Interestingly, the more the patient feels like she is participating in the decision-making, perhaps the less vulnerable the physician may be to liability, and the less vulnerable the physician is to burnout and decreased job satisfaction [51]. A very Flexnerian trajectory has led to the reporting of medical errors amidst an evolving health care culture that values self-correction. One small vehicle of that process is patient/physician joint participation in medical decision-making.

1.8 Lifelong Learning and Continuing Medical Education: Flexner's Ideals Applied to Physicians Throughout Their Careers

Interdisciplinary research is at the foundation of what the Flexner Report proposes as the future of clinical practice and informed medical decision making; interdisciplinary research and medical care is at the heart of women's health care. Lifelong learning essential for this endeavor.

Abraham Flexner did not actually use the term lifelong learning, but it is implicit in everything for which he stood. For physicians to provide high quality and current medical care to patients, they must be continuously educated, finding out the latest features of medicine relevant to their practice [52]. Formal processes of continuing medical education (CME) support the universal endeavor to stay informed, and are essential to the practice of medicine. For the physician, CME lies at the heart of the evolving medical decision making process where decision-making takes into account new innovations and scientific studies. CME helps physicians keep up with the new medical findings and technologies that potentially impact practice. Evidence-based practice guidelines developed by CME societies are critical for continued physician education.[21]

Core features of CME are:

(1) CME planning and program development should be based on needs assessment, including outcome data.
(2) The goals of CME should include the development of skills necessary for lifelong learning, the exercise of clinical reasoning, an understanding of the decision-making process, and specific content acquisition.
(3) The multiple goals of CME should be reinforced by the appropriate choice of learning methods.
(4) Incorporation of new instructional technologies for CME should be based on their intrinsic strengths as learning tools after thorough evaluation.
(5) Faculty development is important within CME and should include exposure to new learning methods (theory and application) enabling faculty to translate their content expertise into formats more appropriate to learners' needs.
(6) Educational activities should b supportive of and coordinated with the transition to evidence-based medicine.
(7) Professional and, whenever possible, interdisciplinary interaction should be given priority in CME programming.
(8) Outcomes-based measures of CME effectiveness and research should be introduced into the determinants of physicians' practice behaviors.

Perhaps most importantly, the goals of CME include assisting healthcare providers with developing lifelong learning skills, exercising clinical reasoning, optimizing physician-patient interaction, and understanding the decision making process, as well as providing specific content to learn. Traditional CME programs have addressed the education and training needs of doctors from residency through retirement and often focus on expanding a physician's knowledge without instructing him or her about how to use that knowledge to solve clinical problems.

In the past 10 years CME programs have begun to focus on EBM and have tried to incorporate new information into a clinical context and link learning to public health outcomes [53]. Self-assessment, problem solving, and evidence-based analysis are three educational techniques that have slowly become topics for CME over the past 30 years. In reality these three techniques may be more helpful than purely science-content driven CME, as previous research has shown that physicians who are exposed to and engage in self-directed, self-reflective learning processes tend to keep up-to-date later in practice.[22]

CME's central objective should be to help the learner recognize what he or she still needs to know. This self-assessment function encourages critical thinking and reinforces the goal of lifelong learning. Physicians benefit from developing techniques that enhance their capacity to evaluate critically their own knowledge and skills and to fill in the gaps where necessary. Decision analysis is an important tool to enhance physicians' self-reflection about biases and orientation to events. Decision analysis entails, in part, a formal scientific approach to decision-making, in which the value or utility of each of decision's possible consequences is multiplied by the probability of the consequence, thus arriving at a score for each outcome.[23]

While a good deal of self-directed learning may be independent learning, not all learning should occur independently. CME appears most likely to affect outcomes when program content is discussed and reinforced in a group. Problem-based learning (PBL) also holds several advantages as a CME method; it encourages self-directed learning within the context of a group or team using a case-management discussion. Incidentally, the dilemmas that pervade hormone therapy provide an ideal context for PBL. PBL nurtures and exercises an assortment of skills that go beyond knowledge acquisition yet are essential to learning and clinical proficiency such as self-assessment, information management, the exercise of clinical reasoning, and a collaborative team-management approach [54]. Continuing medical education courses that encourage evidence-based assessment as it relates to clinical management provide an important educational vehicle for busy physicians [55].

Practice behaviors vary and these variations have led to an emphasis on developing standards for quality of care that are evidence-based, both nationally and within large healthcare delivery organizations. Typically, efforts to promote such changes (i.e., development of standardized clinical protocols, guidelines, and other quality-assurance mechanisms) also signal the need for CME. Therefore, CME should be undertaken in concert with the development and implementation of evidence-based clinical standards and guidelines, and CME program content and formats should complement these goals.

Much of the variation in physicians' practice styles may depend on medical culture. To be effective within an environment where medical culture varies, CME would have to focus more on changing the habits and attitudes of physicians and altering cultural expectations than on strictly providing information. Information alone would be unlikely to have an effect, and might even be discounted if it were not consistent with physicians' preexisting beliefs. Consider, for instance, estrogen therapy. Physician responses to the Women's Health Initiative Study show how significantly prior beliefs, practice, and training affect how physicians evaluate and use new information. Additional research exploring these discrepant responses might help determine the motivating factors behind physicians' behaviors and practice styles. Gender, personal experiences and age, all may impact physician readiness, for instance to prescribe hormone therapy, as well as physician responses to other changes in medical information [56].[24]

1.9 CME and Avoidance of Physician Burnout

In an age of significant physician burnout where physicians have less control, have an ever expanding body of information to tackle, and, most importantly, experience increased lawsuits and insurance costs, physicians, at least in obstetrics and gynecology, have been leaving their profession in record numbers [57]. Job dissatisfaction has become an increasing problem for the medical professions. Liability insurance has been reported as the most significant factor contributing to physicians leaving the field of obstetrics and gynecology and other fields of medicine.[29] Addressing physician burnout could improve both the field of obstetrics and gynecology and care that is delivered to patients. Studies have shown that one way to lessen physician burnout is to increase opportunities to attend CME.

In one study by Bettes et al., [58] the greater the interest of OB-GYNs in CME and, presumably, continuous learning, the greater the decrease in reported physician burnout. This suggests that obstetrician-gynecologists view CME positively. Amidst the diverse conflicts and limitations that make the practice of medicine exceedingly difficult, one important source of comfort is a good dose of education through CME.

1.10 Changing Landscapes: Part-Time Options, Lifestyles, and Physician Satisfaction

One major change in the culture of medicine, especially within obstetrics and gynecology, is the increasing number of women in the medical field. The changing demographics of physicians, along with part-time work options and innovative ways of delivering medical care to patients, perhaps provides more balance to medicine and allows more women to participate in the field of women's health.

As mentioned above, there were few women in medicine before Flexner's report and even fewer after it. Although the Flexner Report was supportive of women in medicine, Flexner was mistaken when he suggested that medical schools at that time were "open to women in practically the same terms as men;" [25] that would take many many more years.

Indeed, most contemporaries of Flexner, such as Walter Channing, an obstetrician and professor at Harvard in the 19th century, believed that women did not have the physical, mental, or emotional capacity to withstand medical school. Fortunately this sentiment is long since forgotten. Indeed, in US obstetrics and gynecology residency programs over 78% of trainees are women.[26]

For the first two-thirds of the 20th century, women pursuing medical education were few and far between. In 1965, women only made up 9% of accepted applicants to medical school. This drastically changed over the following 4 decades. The 2008–2009 AAMC survey for the *Women in the U.S. Academic Medicine Statistics and Medical School Benchmarking Report* found that 48% of students accepted to medical school and 45% of residents were women. Specifically in obstetrics and gynecology, 78% of residents in 2008 were women [59].

Looking specifically at female residents and their career choices, one study found that first year female obstetrics-gynecology residents were more inclined to go into research than first year men, but this difference declined by the fourth year. This change may be due to the perception of the mentoring process and the actual mentoring process of potential clinician-researchers. A Flexnerian orientation aims at making both research and academics widely accessible to all trainees and physicians.[27]

Also emerging in academic medicine and obstetrics and gynecology is the increasing interest in part-time positions. As of 2006, 1% of men and 18% of women academic physicians under the age of 50 years were working part time [60]. The possibility of part-time work or more flexible work options, though also important to men, may be especially important to women because of continued competing family roles. Despite working the same number of hours as their male counterparts, women at a single medical institution reported dedicating 12 h more per week to household and family duties ("the second shift") [61]. Contributing to this difference is likely the fact that 70% of women faculty have partners who work full time, compared to 36% of male faculty.

Trying to accommodate both men's and women's work environment needs in obstetrics and gynecology training is necessary for optimizing education and satisfaction and reducing burnout. A source of significant stress for students and trainees is the conflict between training responsibilities and family responsibilities. It is unlikely that this stress "ends" after residency, though increased autonomy after training may reduce its levels. Just as "patients" jumped to the forefront in shared-medical decision making, "physicians" as individuals with personal needs and out-of-work responsibilities have become a factor in the health care delivery system. As our medical system evolves, our profession must consider how to allow for flexible training and flexible career options to attract the best possible scholars into medical fields and to keep well-trained physicians satisfied so that they will not leave medicine.

The positive job satisfaction rates of part-time ob-gyns probably reflects the increasing importance of a flexible lifestyle.[28] Career satisfaction depends on practice environment, practice autonomy, and market factors. A study looking at how gender affects job satisfaction among obstetricians and gynecologists found that men were more likely to be working full time (91% men vs. 76% women) and that women were more likely to reduce working hours and salary because of parenting. It also found that although women and men rated job satisfaction no differently, women were less likely to indicate that they would not choose a career in obstetrics and gynecology again (19% women vs. 35% men). Physicians in our study were more satisfied when they were working closer to their desired number of hours per week and could balance their work and personal lives (Fig. 1.3).

Since 2005, a dozen states and more than 15 specialties have reported a physician shortage or anticipate one in the next few years. Dissatisfaction with medicine as a career is associated with burnout and the desire to retire early. Dissatisfied physicians have been found to be 2–3 times more likely to leave medicine than satisfied physicians. Obstetrics and gynecology is especially vulnerable to experience this type of shortage. High levels of dissatisfaction, caused in part by increased liability, are thought to be a major factor in early retirement of ob-gyns.

Further,[29] we examined how career satisfaction related to part-time work, increasing liability, current retirement plans, and professional concerns. The results

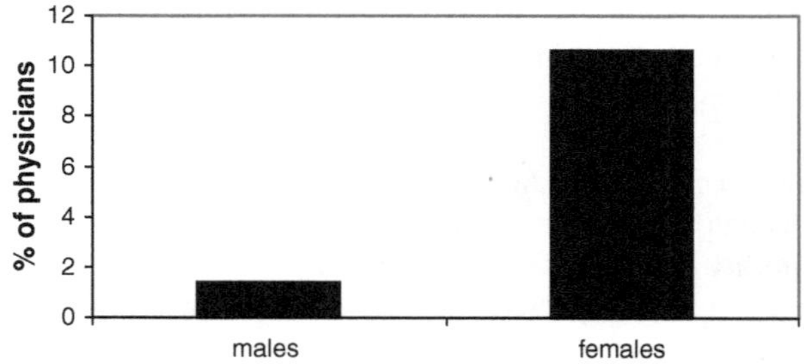

Fig. 1.3 Percent of male and female physicians under age 50 who are working part-time was significantly different

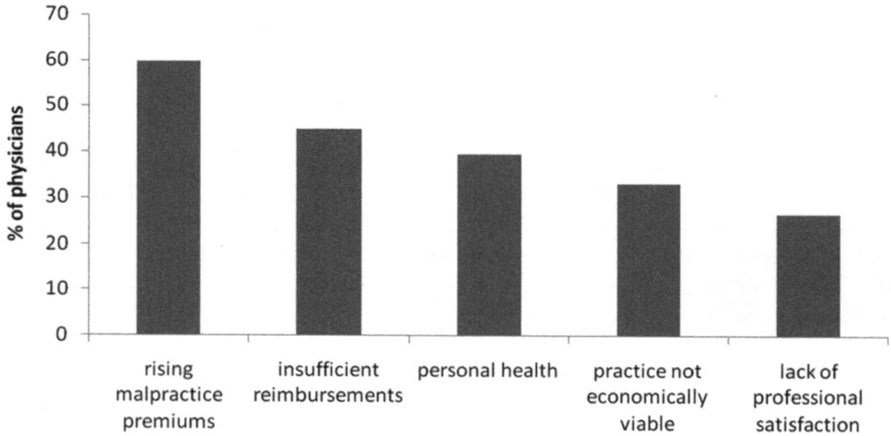

Fig. 1.4 Reasons that obstetrician-gynecologists over the age of 50 years retired earlier than expected

suggested that ob-gyns over age 50 years were more satisfied if they had the option to work part time, if they were less concerned about rising malpractice costs, if they were able to balance their work and personal life, and if they were able to control work hours/schedule (Fig. 1.4).

An entity not considered whatsoever in Flexner's time was the overwhelming impact of medical malpractice and its affect on career longevity and job satisfaction. Malpractice costs have been reported to be an important factor when planning for retirement. Nearly 2/3 of ob-gyns over age 50 reported retiring early, at least in part, because of rising malpractice premiums. Reduced accessibility to women's health care is likely not only because of the increasing demand for medical services and an aging physician population but also, as the study suggests, because dissatisfied ob-gyns are more likely to retire early and new ob-gyns work fewer hours than ob-gyns in the past.

1.11 Conclusion: Flexner for Any Century

The Flexnerian ideal merged the delivery of patient-focused care with a culture of research in medical schools and was zealously incorporated into medical education and medicine in general in the first part of the 20th century. The Flexner Report provided the impetus for medical education to reform and evolve into its modern state, incorporating science into medical education and practice. The Flexner Report was part of a cultural revolution in which scientific discovery and sensibility were codified, expanded, and embedded into the fabric of medical education. Decision science and evidence-based medicine were trajectories off of Flexner's basic approach.

Evidence-based medicine has been embraced at the core of modern medical decision-making. A clear understanding of the cognitive basis of human information processing paired with decision-science and evidence-based medicine enhances the credibility of medical decisions. Almost 50 years ago, obstetricians and gynecologists were noted to be the medical professionals least likely to incorporate evidence from clinical trials into their clinical practice. Medicine and obstetrics and gynecology have undergone a major cultural shift and have embraced evidence-based approaches to medical decision making and patient care. Survey after survey from the American College of Obstetricians and Gynecologists have confirmed that obstetricians-gynecologists regularly review the organization's evidence-based practice guidelines and rely upon them to assist with the delivery of evidence-based health care.

Flexner emphasized enhancing the practice of medicine with the culture of research. He wanted medical schools and hospitals to be held accountable for the physicians they produced and let practice. Flexner's goals included the scientific discernment of physician responsibility, public health, management of health, and care of individuals. Evidence-based medicine furthers those goals by involving collective decision-making in evaluating and grading evidence. Decision science emphasizes individual choice amidst human capabilities and information processing. Both of these evolving themes in medicine are touted within Flexner's conception of cognitive cultivation in the practice of medicine.

Lastly, moving from the "practice of medicine" to the "practitioner" of medicine, we have seen a changing landscape in terms of "who" is pursuing careers in medicine and "what" these careers look like. Physicians are increasingly looking for "part-time" options and professional societies and employers are recognizing that allowing physicians more control has potential advantages of increased job satisfaction and decreased burn-out. Participation in CME and lifelong learning are necessary for all medical practices and healthcare practitioners, and actually may facilitate improved career satisfaction.

We have come a long way from trade schools as a means of medical education to the intricate incorporation of research and evidence into the culture of medicine. The medical field has changed drastically over the past 100 years to include women and minorities, to standardize medical education, and to incorporate clinical research and evidence into a clinical practice that relies upon the patient as a decision-making partner in the clinical encounter. Despite these dramatic advances, the medical field must continue to critically evaluate how it is educating its students and delivering its medical care so that advances in medicine and patient care can be continued into the future.

Notes

1. See Gigerenzer [1]
2. Atran [6]
3. Oshinsky [14]
4. See Flexner [4]
5. See Dewey [1]
6. See *supra* note 1; See Abrahamson [21]
7. See Ludmerer [9]
8. See Cook [26]
9. See Cook et al. [23]
10. See Dickersin and Menheimer [26]
11. Morgan et al. [35]
12. See Baron [38]
13. See *supra* notes 1 and 12
14. See Gilovich et al. [39]
15. See *supra* note 14
16. See Elstein [36]
17. See Faden and Beauchamp [44]; see Beauchamp and Childress [44]
18. See Pellegrino and Thomasm [45]; see Veatch [46]
19. See Moore et al. [48]
20. See Veatch, *supra* note 18
21. See Abrahamson, *supra* note 6
22. *ibid*
23. See *supra* note 16; see Schulkin [36]
24. See Lyerly [46]
25. See *supra* note 4
26. See Cain [17]
27. *Ibid*
28. See Anderson [60]
29. *Ibid*

References

1. Schulkin, J. 2009. *Cognitive adaptation: A pragmatist perspective*. Cambridge: Cambridge University Press; Gigerenzer, G. 2009. Making sense of health statistics. *Bulletin of the World Health Organization* 87:567; Gigerenzer, G. 2000. *Adaptive thinking: Rationality in the real world*. New York, NY: Oxford University Press; Dewey, J. 1925/1989. *Experience and nature*. La Salle: Open Court; Dewey, J. 1938/1972. *Logic: The theory of inquiry*. New York, NY: Henry Holt.
2. Foley, R. 1996. An evolutionary and chronological framework for human social behavior. *Proceedings of the British Academy* 88:95–117.
3. Findley, P. 1939. *The priests of Lucina: The story of obstetrics*. Boston, MA: Little, Brown.
4. Flexner, A. 1910. Medical education in the United States and Canada: A report to the carnegie foundation for the advancement of teaching. *Bulletin Number Four*, New York City, 2.
5. Schulkin, J. 2004. *Bodily sensibility: Intelligent action*. Oxford: Oxford University Press, 3.

6. Atran, S. 1990/1996. *Cognitive foundations of natural history.* New York, NY: Cambridge University Press.
7. Bynum, W. 2008. *The history of medicine: A very short introduction.* Oxford: Oxford University Press.
8. Starr, P. 1982. *The social transformation of American Medicine.* New York, NY: Basic Books.
9. Ludmerer, K. 1996. *Learning to heal: The development of American medical education.* Baltimore, MD: The Johns Hopkins University Press.
10. Drife, J. 2002. The start of life: A history of obstetrics. *BMJ* 78:311–315.
11. Bernard, C. 1865/1957. *An introduction to the study of experimental medicine.* New York, NY: Dover Press.
12. Osler, W. 1892. *Principles and practices of medicine.* New York, NY: D. Appleton and Company; Bliss, M. 1999. *William Osler: A life in medicine.* Oxford: Oxford University Press.
13. Longo, L.D. 1980. *Obstetricts and gynecology: The education of American physicians,* ed. R.L. Numbers, pp. 66–77. Berkeley, CA: University of California Press, Kelly.
14. Oshinsky, D.M. 2005. *Polio: An American story.* Oxford: Oxford University Press.
15. Bonner, T.N. 2002. *Iconclast: Abraham Flexner and a life in learning.* Baltimore, MD: Johns Hopkins University Press.
16. Hiatt, M.D., and C.G. Stockton. 2003. The impact of the Flexner report on the fate of medical schools in North America after 1909. *Journal of American Physicians and Surgeons* 8(2): 37–40.
17. Walsh, M.R. 1992. Women in medicine since Flexner. In *Beyond Flexner: Medical education in the twentieth century*, eds. B. Barzanksy and N. Gevitz, pp. 82–92. New York, NY: Greenwood Press.
18. Cain, J.M. et al. 2001. Effects of perceptions and mentorship on pursuing a career in academic medicine in obstetrics and gynecology. *Academic Medicine* 76:628–634; Coleman, V.H. et al. Obstetrician-gynecologists and primary care: Training during obstetrics-gynecology residency and current practice patterns. *Academic Medicine* 82:602–607.
19. Hoover, E.L. 2006. Did Flexner's Report condemn black medical schools? Not so, in my opinion. *Journal of the National Medical Association* 98(9):1432–4; Savitt, T.L. 1992. Abraham Flexner and the black medical schools, in Beyond Flexner: Medical education in the twentieth century, eds. B. Barzansky, and N. Gevitz. New York, NY: Greenwood Press; Steinecke, A., and C. Terell. 2010. Progress for whose future? The impact of the Flexner report on medical education for racial and ethnic minority physicians in the United States. *Academic Medicine* 85:236–245.
20. Gross, J.P. et al. 2008. After a century of criticizing premedical education, are we missing the point? *Academic Medicine* 83:516–520; Thomas, E. 2001. Empathy and conciousness. *Journal of Consciousness Studies* 8:1–35.
21. Abrahamson, S. et al. 1999. Continuing medical education for life: Eight principles. *Academic Medicine* 74:1288–1294; Miller, B.M. et al. 2010. Beyond Flexner: A new model for continuous learning in the health professions. *Academic Medicine* 85:266–272.
22. Gabbe, S.G. 1996. After office hours: The alphabet of academic medicine. *Obstetrics and Gynecology* 88:479–481; Gabbe, S.G. 1998. Graduate education: A blueprint for academic obstetrics and gynecology. *Obstetrics and Gynecology* 92:1033–1037.
23. Cooke, M., D.M. Irby, W. Sullivan, and K.M. Ludmerer. 2006. American medical education 100 years after the Flexner report. *New England Journal of Medicine* 355(13):1339–44; Schulkin, J. 2000. Decision sciences and evidence-based medicine: Two intellectual movements to support clinical decision-making. *Academic Medicine* 75:36–38; Morrison, G., S. Goldfarb, and P.N. Lanken. 2010. Team training of medical students in the 21st century: Would Flexner approve? *Academic Medicine* 85:254–259; Skochelak, S.E. 2010. A century of progress in medical education: What about the next 10 years? *Academic Medicine* 85:197–200; Kirch, D.G. 2010. The Flexnarian legacy in the 21st century. *Academic Medicine* 85:190–192.

24. Irby, D.M., M. Cook, and B. O'brien. 2010. Calls for reform of medical education by the carnegie foundation for the advancement of teaching: 1910 and 2010. *Academic Medicine* 85:220–227.
25. Sackett, D.L. et al. 1997. On the need for evidence-based medicine. *American College of Physicians* 1:5–9.
26. Dickersin, K., and E. Manheimer. 1998. The cochrane collaboration: Evaluation of healthcare and services using systematic reviews of the results of randomized controlled trials. *Clinical Obstetrics and Gynecology* 41:315–331.
27. National Perinatal Epidemiology Unit. 1985. *A classified bibliography of controlled trials in perinatal medicine: 1940–1984*. Oxford: Oxford University Press.
28. Chalmers, I., L.V. Hedges, and H. Cooper. 2002. A brief history of research synthesis. *Evaluation and the Health Professions* 25:12–37.
29. Cochrane, A.L. 1972. *Effectiveness and efficiency*. London: Nuffield Provincial Trust; Jadad, A.R., and R.B. Haynes. 1998. The cochrane collaboration-advances and challenges in improving evidence-based medicine. *Medical Decision Making* 18:2–9.
30. Sackett, D.L. 1994. The cochrane collaboration. *American College of Physicians* 120:1–11.
31. Daly, J. 2005. *Evidence-based medicine and the search for a science of clinical care*. Berkeley, CA: University of California Press.
32. Grimes, D.A. 1995. Introducing evidence-based medicine into a department of obstetrics and gynecology. *Obstetrics and Gynecology* 86:451–457.
33. Olatonbosum, O.A., L. Edouard, and R.A. Pierson. 1998. Physician attitudes toward evidence-based obstetric practice. *BMJ* 316:365–366.
34. Farquhar, C.M. et al. 2002. Clinical practice guidelines as educational tools for obstetrician-gynecologists. *Journal of Reproductive Medicine* 47:897–902.
35. Morgan, M.A. et al. 2005. Impact of self-reported familiarity with guidelines for cystic fibrosis carrier screening. *Obstetrics and Gynecology* 61:1355–1361.
36. Elstein, A.S. 1989. Decision analysis in surgical education. *World Journal of Surgery* 13:287–291; Schulkin, J. 2000. Decision sciences and evidence-based medicine-two intellectual movements to support clinical decision making. *Academic Medicine* 75(8):816–818.
37. Timmermans, S., and A. Macuk. 2005. The promise and pitfalls and evidence-based medicine. *Academy of Medical Affairs* 24:18–28; Muir-Gray, J.A. 1997. *Evidence-based healthcare*. New York, NY: Churchill Livingston.
38. Baron, J. 2000. *Thinking and deciding*. New York, NY: Cambridge University Press.
39. Gilovich, T., D. Griffin, and D. Kahneman, eds. 2002. *Heuristics and biases: The psychology of intuitive judgment*. New York, NY: Cambridge University Press.
40. Kahneman, D., P. Slovic, and A. Tversky, eds. 1982. *Judgement under uncertainty: Heuristics and biases*. New York, NY: Cambridge University Press.
41. Baron, J., C.B. Holzman, and J. Schulkin. 1998. Attitudes of obstetricians and gynecologists toward hormone replacement therapy. *Medical Decision Making* 18:406–411; Power, M.L., J. Baron, and J. Schulkin. 2008. Factors associated with obstetrician-gynecologists' response to the women's health initiative trial of combined hormone therapy. *Journal of Reproductive Medicine* 51:525–532.
42. Schkade, D.A., and D. Kahneman. 1998. Does living in California make people happy? *Psychological Science* 9:340–346.
43. O'Connor, A.M. et al. 2005. Decision aids for people facing health treatment or screening decisions. *Cohcraine Databse of Systematic Reviews* 4:113–119.
44. Faden, R.R., and T.L. Beauchamp. 1986. *A history and theory of informed consent*. Oxford: Oxford University Press; Beauchamp, T.L., and J.F Childress. 2001. *Principles of biomedical ethics,* 5th edn. Oxford: Oxford University Press.
45. Pellegrino, E.D., and D.C. Thomasma. 1998. *For the patient's good.* Oxford: Oxford University Press.

46. Veatch, R.M. 1981. *A theory of medical ethics*. New York, NY: Basic Books; Lyerly, A.D., E.R. Myers, and R.R. Faden. 2001. The ethics of aggregation and hormone replacement therapy. *Health Care Analysis* 9:187–211.
47. Patient Testing: Ethical Issues in Selection and Counseling. ACOG Committee Opinion Number 363, 2007.
48. Moore, D.A. et al. 2005. *Conflicts of interest*. Cambridge: Cambridge University Press.
49. Gert, B. et al. 1997/2006. *Bioethics: A return to fundamentals*. Oxford: Oxford University Press.
50. Warner, R. 1980. *Morality in medicine*. Sherman Oaks, CA: Alfred Publishing.
51. Bettes, B.A., A.L. Strunk, V.H. Coleman, and J. Schulkin. 2004. Professional liability and other career pressures: Impact on obstetrician-gynecologists' career satisfaction. *Obstetrics and Gynecology* 103:967–973.
52. Sullivan, W. 2010. Expanding Flexner's legacy through new understanding. *Academic Medicine* 85:201–202; Maeshiro, R. et al. 2010. Medical education for a healthier population: Reflections on the Flexner report from a public health perspective. *Academic Medicine* 85:211–219.
53. Davis, D.A., and M.A. Thomson. 1996. Implications for undergraduate and graduate education derived from quantitative research in continuing medical education: Lessons learned from an automobile. *Journal of Continuing Education in the Health Professions* 16:159–166.
54. Schmidt, H.G. 1993. Foundation of problem-based learning: Some explanatory notes. *Medical Education* 27:422–432; Lewis, C.E. 1998. Continuing medical education: Past, present, future. *Western Journal of Medicine* 168:334–340.
55. Henderson, V.W. 2004. Hormone therapy and alzheimer disease: An evidence-based approach to clinical management. *Female Patient* 29:17–22.
56. Biglia, N. et al. 2004. Personal use of HRT by postmenopausal women doctors and doctor's wives in the north of Italy. *Gynecological Endocrinology* 18:165–174.
57. Williams, E.S., and A.C. Skinner. 2003.Outcomes of physician job satisfaction. *Health Care Review* 38:119–139; Mello, M.M., D.M. Studdert, and T.A. Breman. 2003. The new medical malpractice crisis. *New England Journal of Medicine* 99:2281–2284.
58. Bettes, B.A. et al. 2004. Heavier workload, less personal control: Impact of delivery on obstetrician-gynecologists' career satisfaction. *American Journal of Obstetrics and Gynecology* 190:831–837; Bettes, B.A. et al. 2004. Professional liability and other career pressures: Impact on obstetrician-gynecologists' career satisfaction. *Obstetrics and Gynecology* 103:967–973.
59. Association of Medical Colleges (AAMC). Women in U.S. Academic Medicine Statistics and Medical School Benchmarking, 2007–2008.
60. Anderson, B.L., R.W. Hale, E. Salsberg, and J. Schulkin. 2009. Outlook for the future of the obstetrician-gynecologist workforce. *American Journal of Obstetrics and Gynecology* 201(4):e12, Oct 2009.
61. Shollen, S.L., C.J. Bland, D.A. Finstad, and A.L. Taylor. 2009. Organizational climate and family life: how these factors affect the status of women faculty at one medical school. *Academic Medicine* 84(1):87–94.

Chapter 2
Generational and Gender Shifts in Academic Medicine

Breda Bova and Sharon Phelan

Abstract Many examples of generational and gender shifts are noted in academic medicine, perhaps most notably in women's health care. This chapter describes trends in faculty demographics, size, and responsibilities and projects future directions in growth.

Keywords Work-life · Generation · Ob-gyn · Workforce

2.1 Introduction

According to Schaffner and Van Horn, "the workplace is our most diverse national institution, and its diversity will only grow" [1]. This includes ethnic, racial and gender differences but increasingly, examining and understanding generational diversity has become an important part of maximizing organizational effectiveness.

Two striking demographic shifts evident in our national workforce today are also apparent in the medical profession. One is the large number of women in U.S. graduate and medical schools. The other is the new generations of physicians, both Gen Xers and Gen Y. Both of these shifts have significant implications for academic medicine in terms of recruitment and retention. This is due to the attitudes of the younger generations in regards to career development and work- life balance. A desire for improved work-life balance has prompted more physicians, including Ob-Gyns, to want to work fewer hours [2]. In addition, America is aging, resulting in a large number of retirements expected in all professions over the next 10 years. Medicine is no different.

In their book, The Coming Generational Storm, authors Kotlikoff and Burns [3] comment,

B. Bova (✉)
1 University of New Mexico, Albuquerque, NM 87131, USA
e-mail: bova@unm.edu

W.F. Rayburn, J. Schulkin (eds.), *Changing Landscape of Academic Women's Health Care in the United States*, International Library of Ethics, Law, and the New Medicine 48,
DOI 10.1007/978-94-007-0931-7_2, © Springer Science+Business Media B.V. 2011

> The aging of America isn't a temporary event. We won't be getting older this year or this decade, and then turning back and getting younger. We are well into a change that is permanent, irreversible, and very long term. Where we had 35.5 million people age 65 and older in 2000, we'll have 69.4 in 2030.

Educational shifts are also affecting the workplace. According to the U.S. Department of Education [4], women have been earning more bachelor's degrees than men since 1982 and more master's degrees than men since 1981.

- In the 2005–2006 academic year (the most recent year for which data are available), women earned 58% of all bachelor's degrees and 60% of master's degrees.
- By 2016, women are projected to earn 60% of bachelor's, 63% of master's and 54% of doctorate and professional degrees.

There is no question that our workforce and workplace is changing. This article examines the impact of generational issues, gender shifts in the workforce, the implications of more people wanting to work part time and the effect that work- life balance is having on all academic departments but particularly Ob-Gyn and Family Medicine departments which are a the major training programs for providers of women's health.

2.2 Generational Issues

Each generation has its own unique stories, characteristics, differences and tensions. A generation is basically defined as a group of people who share the same formative experiences. These experiences bind people who are born in continuous years into cohorts. These are groups of individuals customarily sorted by birth years as the common demographic statistic. Depending on which demographer you read there may be slight variations in the birth years for each generation as well as different names used for the identification of each. The four generations present in the workplace today are outlined in Table 2.1.

Table 2.1 Four generations in the workplace today

The four generations that remain in the workplace today are
• The Silent Generation (1925–1942) Approximately 63 million
• The Baby Boomer Generation (1943–1961) Approximately 77 million
• Generation X (1962–1981) Approximately 44 million
• Generation Y (1982–2000) Approximately 70 million

2.3 The Silent Generation (1925–1942)

Also known as Radio Babies, Veteran Generation, WW II Generation, Seniors and Geezers. Most of our senior physicians are members of the Silent Generation. Their youth was very different than what young people experience today. They received most of their news from radio and newspapers, shopping was done most often at local shops and markets and long distance phone calls were rare and expensive. Some of the seminal events that influenced this group were the Great Depression, the New Deal, World War II, the rise of labor unions, and the Korean War. Even though the younger members of this generation do not remember the Great Depression, they remember the often severe impact it had on their parents.

Members of this generation were prepared to endure difficult situations, master a body of knowledge, and demonstrate respect for those who came before them. Age and experience counted for something. Members of this generation have adapted to the changing world around them but they continue to rely on the early lessons they learned in the workplace. It's interesting to reflect back on what those early workplaces looked like: heavily bureaucratic, with clearly defined rules, policies and procedures. In most cases some one person was clearly in charge and those who received rewards were usually those who followed the rules. According to Moody, most physicians in this generation grew up in single- income families in which the mother was the homemaker [5]. Some of the characteristics that define this group are shown in Table 2.2.

Table 2.2 The silent generation (Adapted from Sotile W and Bova B. Among the generations in medicine:learning to address impairment and fatigue to enhance patient safety. Duke University NC 2006)

Silent Generation: Characteristics
•Able to postpone Gratification
•Risk Aversive
•Loyal
– Family
– Country
– Job
•Respectful Communication
•Adherence to Rules
•Detail Oriented

2.4 Baby Boomer Generation (1943–1961)

Also know as Digital Immigrants and the Vietnam Generation. This is the largest generation in history. In many workplaces today, including academic medicine, Baby Boomers hold positions of power and control. The youth of this generation grew up with parents who had been through World War II and thought that life would be better for their children. They grew up in a generally prosperous environment and were the first generation to be introduced to television. They became used

Table 2.3 Seminal events of the baby boomers (citation)

Some seminal events affecting the baby boomers	
1955	Salk Vaccine tested on the public
	Rosa Parks refuses to move to the back of the bus in Montgomery, AL
1957	First nuclear Power Plant
	Congress passes the Civil Rights Act
1960	Birth control pills introduced
1961	Kennedy establishes Peace Corps
1962	Cuban Missile Crisis
	John Glenn circles the earth
1963	Martin Luther King leads march on Washington, D.C.
	President John Kennedy assassinated
1965	United States sends ground combat troops to Vietnam
1965	National Organization for Women founded
1968	Martin Luther King and Robert F. Kennedy Assassinated
1969	First Lunar landing
	Woodstock

to getting their news on TV as well as seeing historic events take place such as men landing on the moon and the assassination of a president. Numerous seminal events had an impact on their lives. Some of these are highlighted in Table 2.3.

Baby Boomers are still a powerful presence in the workplace and are likely to dominate it until 2015. Many members of this generation chose the workplace as a vehicle for proving their sense of self worth. Many consider this generation as "driven" – working long hours, missing many of their children's "firsts" and doing whatever it took to get the job done. For many baby boomers this was the only way to the top. They were willing and expected to climb the "ladder" to get there and wait for their turn. This is a phenomenon that is being challenged today. Some of the characteristics that define this group are shown in Table 2.4.

Table 2.4 Characteristics of the baby boomers (Adapted from Sotile W and Bova B. Among the generations in medicine:learning to address impairment and fatigue to enhance patient safety. Duke University NC 2006)

Baby boomers: characteristics
• Largest Generation: 77 Million
• Optimistic
• Redefined Roles
• Management by Buzz Word
• Skewed Work/Life Balance
• Brought Up in a Competitive Environment
• Will transform Retirement models
• "Work Ethic" and "Worth Ethic" are Synonymous

2.5 Generation X (1962–1981)

Also known as Baby Busters, Thirteenth Generation (since the American Revolution) 13ers, and Post Boomers. Generation X is a much smaller group than the Baby Boomers, approximately 44 million persons in the U.S. During their childhood they grew up in homes where many of their mothers worked outside the home.

Table 2.5 Seminal events of generation X (Adapted from Sotile W and Bova B. Among the generations in medicine:learning to address impairment and fatigue to enhance patient safety. Duke University NC 2006)

Some seminal events affecting generation X	
1971	Intel's first chip developed
1972	First e-mail management program
1973	Roe v Wade
1975	Personal computer introduced on the consumer market
1981	Centers for Disease Control's first published report on AIDS
1981	Reagan assassination attempt
1984	Extensive corporate downsizing begins
1986	Challenger explosion

This generation was affected by divorce as many grew up in single parent homes. They became latch key children and took care of their brothers and sisters. This has manifested itself in a generation that is generally self reliant and resourceful. Seminal events that have impacted this group are highlighted in Table 2.5.

Since many in this generation witnessed their parents being cut from jobs during the extensive corporate downsizing that began in the mid 1980s, they tend to have a free agent approach to their careers. This simply means that they, not a profession or institution, own their careers. There is an increase in the number of stay at home Gen X moms, most pronounced among college graduates. They indicate they are looking for a less "frazzled" life style. As one female Gen Xer said, "I saw my parents old and tired at 50 – I can always get another job. I don't want to get another family." In many of our medical institutions this generation makes up our residents and junior faculty.

Characteristics of this generation are detailed in Table 2.6.

This generation introduced us to the concept of work-life balance and in turn this has had a huge impact on the workplace today. Many in this generation truly value their family time and as Bickel and Brown note [6] "in their eyes their parents suffer

Table 2.6 Characteristics of generation X (Adapted from Sotile W and Bova B. Among the generations in medicine:learning to address impairment and fatigue to enhance patient safety. Duke University NC 2006)

Generation X characteristics
• Dedicated to people, projects, ideas and tasks, not to longevity and lifetime employment
• Parallel thinkers
• Independent and resourceful
• Accepting of change
• "Want it now!"
• Comfortable with diversity
• Expect a balanced lifestyles
• Free agent approach to careers

from 'vacation deficit disorder'." In this generation we see an increase in the number of physicians who are choosing to work part time.

In a 2007 questionnaire of graduating medical students by the Association of American Medical Colleges, 52% of respondents agreed that a physician's work 'interfere(s) too much with other interests and pursuits,' while even more felt it interfered too much with family relations [7].

2.6 Generation Y (1982–2000)

Also known as Echo Boomers, Millennials, Digital Generation, .com Generation, Digital Natives and the Net Generation. This is the generation in our medical schools and intern programs today as well as those in the undergraduate and graduate pipeline in colleges and universities. The childhood of this generation was greatly impacted by technology and generally closer relationship with their parents. Seminal events that helped shape this generation are shown in Table 2.7

Generation Y has arrived in the workplace and will continue to populate it for decades. They come with a keen sense of and comfort with technology. Since this generation is so techno savvy, many organizations have started reverse mentoring programs in which members of this generation mentor the more senior generations – especially when it comes to technology.

Characteristics of this generation are shown in Table 2.8.

Table 2.7 Seminal events of generation Y (Adapted from Sotile W and Bova B. Among the generations in medicine:learning to address impairment and fatigue to enhance patient safety. Duke University NC 2006)

Some seminal events affecting generation X	
1986	Challenger Explosion
1995	Oklahoma City bombing
1999	Columbine High School massacre
2001	9/11
2005	Hurricane Katrina
2007	Virginia Tech Massacre

Table 2.8 Characteristics of generation Y (Adapted from Sotile W and Bova B. Among the generations in medicine:learning to address impairment and fatigue to enhance patient safety. Duke University NC 2006)

Generation Y characteristics
• E-Learners
• In a state of continuous partial attention
• Used to instant communication
• Many are into "Extreme Sports"
• Expect frequent and/or constant feedback
• Optimistic
• Speed is valued more than attention to nagging detail
• Oriented toward collective action

The Pew Research Center [8] notes some things about this generation that will impact the workplace:

- They are the most ethnically and racially diverse cohort of youth in the nation's history. Among those ages 15–29: 18.5% are Hispanic; 14.2% are black; 4.3% are Asian; 3.2% are mixed race or other; and 59.8%, a record low, are white.
- They are starting out as the most politically progressive age group in modern history. In the 2008 election, Millennials voted for Barack Obama over John McCain by 66–32%, while adults ages 30 and over split their votes 50–49%. In the four decades since the development of Election Day exit polling, this is the largest gap ever seen in presidential election between the votes of those under and over age 30.
- They are the first generation in human history who regard behaviors like tweeting and texting, along with websites like Facebook, YouTube, Google and Wikipedia, not as astonishing innovations of the digital era, but as everyday parts of their social lives and their search for understanding.
- They are more inclined toward trust in institutions than were either of their two predecessor generations – Gen Xers (who are now ages 30–45) and Baby Boomers
- (now ages 46–64) when they were coming of age.

The question that many are asking is how will this new generation of doctors approach the practice of medicine? Kirch suggests some ideas:

- Clearly, job sharing, part-time work, and flexible schedules must become more commonplace for doctors in the years to come.
- Information technology, particularly electronic medical records and emails between physicians and patients, will need to be expanded in order to maintain continuity of care.
- Medical education will need to focus on interdisciplinary training with nurses, pharmacists, and other health professions, as well as a more team-based approach to patient care.
- As this generation of physicians leads by example and shows us that a work-life balance is possible for doctors, they provide us with an excellent opportunity for constructive change in how we practice medicine" [9].

In their book, *Millennials Rising*, Howe and Strauss comment on the importance of understanding generational transitions.

> "Those who pay no attention to generational transitions, or who think generations matter only to the producers and consumers of pop culture, leave themselves with precious little to assess where they, and America, stand in the sweep of history.... Trying to understand the direction of America by looking at the breaking headlines while ignoring generations is like trying to understand the movement of the ocean by looking at the breaking waves while ignoring the tides. Those who disregard generational change have been surprised by the last several turns in the American mood. Those who continue to disregard it will be just as surprised the next time a new decade and a new generation alter the nation's course [10]."

2.7 Gender Issues

In addition to the major shift in academic workforce due to generational changes, there has been a relatively recent change in the number of women in medical careers. This shift became evident in medical schools in the 1970's. Currently there is a balance between the genders at matriculation. Table 2.9 shows the increase from 8.3% of the 1965 entering class being women to the current rate of almost 50%.

Initially, the increasing number of women did not change the culture of many of the specialties. Women often gravitated to more family-friendly specialties with more outpatient responsibilities and "regular" work hours. These included family medicine, pediatrics, pathology and psychiatry.

In the 1980's a shift in specialty choices became evident. Although the traditional primary ambulatory care specialties of Family Medicine and Pediatrics remained popular more women started entering internal medicine subspecialties and obstetrics-gynecology. Table 2.10 shows these trends within selected specialties. For many, ob-gyn was the most supportive surgical specialty for female graduates through the 1990's. If a woman was drawn towards a surgical career, she would be guided to ob-gyn citing "women taking care of women" and to general surgery if a male. During the 1990's ob-gyn residencies were very popular and very competitive while the gender mix shifted to a female dominated specialty. Obviously anyone who wanted to do obstetrics was drawn to ob-gyn. Women who were part of the women empowerment movement aligned themselves with the "Women caring for Women" mantra. They saw ob-gyn residencies as a duty or calling. Finally women who desired to enter a surgical training program commonly felt they had very limited options especially if they were considering having a family. By 2007, over 75% of ob-gyn residents were women which surpassed the traditionally female preferred pediatric residencies with only 73%. (Table 2.10) [11].

In the 1990's as part of the gatekeeper concept for health care reform, more residencies maintained their primary care status. Ob-gyn was part of this movement, with primary ambulatory care becoming a requirement in resident training.

Table 2.9 The number of women and percentage of women compared to all medical students in the areas of matriculation, first year enrollment and graduation from medical school by year

	Matriculants		First year enrollment		Graduates	
Year	# Women	% Women	# Women	% Women	# Women	%Women
1965	799	9.3	731	8.3	524	6.9
1970	1,228	11.0	1,256	11.1	827	9.2
1975	3,511	23.6	3,647	23.8	2,212	16.2
1980	4,757	28.7	4,966	28.9	3,898	24.9
1985	5,520	33.9	5,800	34.2	4,957	30.8
1990	6,153	38.5	6,550	38.8	5,549	36.0
1995	6,941	42.7	7,363	43.2	6,501	40.9
2000	7,472	45.6	7,659	45.9	6,824	43.2
2005	8,239	48.5	8,416	48.4	7,748	48.7

Table 2.10 Distribution of female residents in selected specialty for certain years

Specialty	1993		1997		2003		2007	
	% women residents in specialty	% residents are women	% women residents in specialty	% residents are women	% women residents in specialty	% residents are women	% women residents in specialty	% residents are women
Anesthesiology	4.3	22.9	3.0	26.5	3.2	27.2	4.5	35.1
Dermatology	1.5	51.8	1.3	49.4	1.4	57.5	0.9	62.5
Emergency Medicine	2.0	25.6	2.6	27.6	3.3	32.1	4.4	39.0
Family Medicine	10.2	39.3	13.6	44.9	12.2	51.0	12.8	54.2
Internal Medicine	20.9	31.3	23.1	37.1	22	41.1	24.3	43.9
Ob-gyn	8.7	53.0	8.8	62.6	8.7	74.4	9.1	76.7
Pediatrics	14.4	59.6	15.8	60.6	13.3	68.1	17.8	69.0
Surgery	4.6	17.1	4.6	20.5	4.9	25.4	2.9	30.8

At the same time traditional primary care residencies such as Family Medicine and Internal Medicine added the option of a Women's Health specialty. These programs emphasised the concept that women's health care is not simply a standard approach to care but unique taking into consideration the physiologic differences of women compared to a men. This expanded the concept of women's health is more than reproductive health but examining the unique impact of women's hormonal changes and physiology on their health care.

In 2003, ACGME implemented work hour restrictions. Suddenly the traditional surgical work week of 100–120 h was no more. The work environment between the various surgical specialties became more similar and surgery minded female graduates could started competing for all surgical residency positions not just ob-gyn residency slots [12]. The scope of women's health care is changing. More and more patient care is in an ambulatory setting with fewer surgical or hospital based interventions. The combination of cervical CIS and desire for tubal sterilization used to justify a vaginal hysterectomy. Now it is treated with a LEEP and a laparoscopic or even hysteroscopic procedure. The specialty of Ob-Gyn has the potential of being far more ambulatory in the care provided, making it more similar to family medicine or internal medicine.

It was during this transition that ob-gyn residencies found that their popularity with U.S. medical school graduates had decreased significantly. Unfortunately the two other specialties that might stress Women's Health Care (Family Medicine and Internal Medicine) were not particularly popular either. Issues of part-time employment, comfort with salaried positions rather than private practice, interest in control of work hours and openness to shift work were all concepts foreign to many of the senior leaders in academic medicine, especially in surgical and inpatient specialties. During discussions by various groups, including ACOG, it became evident that this shift in priorities was more than just a gender issue. It was very much a generational issue. Both men and women were expressing the desire for life balance and more control over life style. Specialties seen as lifestyle friendly were those with more shift based practices (Emergency Medicine, Radiology, Anesthesia), or increasingly ambulatory settings with procedures (Dermatology, Otolaryngology). As a group these specialties were commonly referred to as the E-ROAD residencies. They were specialties that were well paid, seen as having status within the medical system and were flexible for part-time or shift arrangements. Women as well as men started entering these competitive specialties in increasing numbers. The newly devised roles of hospitalist, intensivists, and even laborist appear to be more popular among new graduates because of the challenges of the hospitalized patient together with the benefits of limited shift work and salaried employment by the hospital without the administrative headaches of a private office. Primary care and Ob-Gyn were not as popular due to concerns of low pay and/or long hours. Yet these are the groups providing the majority of women-focused health care. Academic departments may find it difficult to effectively recruit new graduates into academics as they compete with private settings that offer a variety of job settings and greater flexibility.

2.8 Work Life Balance

Work-life balance has been cited as one of the most important considerations in choosing a specialty as well as choosing a practice setting. According to the Locum Tenens and the American College of Physicians Executives (ACPE) study, it is noted that:

> More critical today than ever before, supporting physicians' needs for work-life balance can improve their satisfaction on the job. When we fail to recognize the importance of work-life balance for employees, we lose a tremendous opportunity not only to demonstrate understanding and compassion, but to provide appreciable help in what has become a significant challenge for most people [13]

Work-life balance is a difficult concept to define and in many cases one definition does not fit all. In its broadest sense Clark and Hill define it as,

> A satisfactory level of involvement or 'fit' between the multiple roles in a person's life. Although definitions and explanations vary, work/life balance is generally associated with equilibrium, or maintaining an overall sense of harmony in life. This would mean that psychological aspects of work and family life determine balance. Balance is enhanced when individuals engage in doing those things that are salient, prominent, noticeable, or important. Balance is enhanced when individuals choose to act on the things they value [14].

It's a complex concept that Generation X and Y have really elevated to a new level. Many Baby Boomers feel that they are responsible for the work-life balance of everyone but themselves. So in many arenas the question is – who will do the work? The younger two generations watched as their parents worked long hours and as one Gen Xer said "My parents both worked and missed so much of my life as I was growing up. I decided that I would do everything I could to never let that happen to me and my family."

Since turnover and retention are a major issue for the medical field, attention to the issue of what work-life balance means for the younger physician is significant , especially for fields that are critical for public health but currently not popular with the younger generation. There is a need to move away from the notion that those who seek flexibility or part time work are lazy and less committed. Organizations also need to realize that work-life balance has relevance to all generations, not just Gen X and Y. There is no "one size fits all" model. It's important for institutions to offer practices that appeal to all employees. This includes our baby boomers, who have needs in the area of work-life balance as well.

The success of our academic medical organizations depends on the satisfaction of all of our employees. For our medical institutions to succeed our people need to succeed, not just in the workplace but in all aspects of their lives. Here lies the importance of our consideration of the concept of work-life balance. As with any other new Human Resource initiative, it begs us to be open and willing to work with the entire person for the good of the organization as well as the profession.

2.9 Impact of the Changing Workforce on Academic Manpower

Recent resident graduates are more open with their goals in life – both personal and professional. They are striving for a balance as necessary for success at work and at home. An unmanageable schedule and lack of sense of control can lead to depression, poor work performance, conflict with family members, and a sense of burnout. Burnout can in turn lead to leaving the work place or even the profession. Worse those with burnout will stay in the profession simply to earn a salary. Women are often at danger from such an unmanageable schedule because they commonly feel responsible for issues on the home front as well as at work. New graduates need to make tough decisions and accept compromise to regain/maintain control of their lives. Residents often see their faculty as having long hours, many demands on time, and limited to poor work-life balance.

Academic physicians are expected to be productive as clinicians, educators and researchers. To truly be a triple threat, the home life is likely to suffer. The advantages of a salaried position and "regular" hours are often overwhelmed by the demands of clinical productivity in an unwieldy large academic clinic, submitting detailed research grants that are more commonly not funded in today's economy and teaching in one's spare time [15]. Between 1999 and 2003 approximately 15% of graduates entered an academic position with around 40% being women. This contrasts to over 30% going into advance training with approximately32% being women and 48% entering non-academic positions with slightly less than 40% being women [16]. Most of the increases in academic positions were clinical positions [17]. To keep this faculty, academic leadership will have to address the individual pursuit for balance and alignment in work and at home [18]. Gender does impact faculty job satisfaction and men are generally being more satisfied [19]. This makes Obstetric and Gynecology and Pediatric departments especially vulnerable to problems with maintaining an adequate academic workforce since most of the younger graduates are women.

Interest in part time employment/careers is increasing to achieve this balance. Although this sounds simple on the surface to simply have every Wednesday off to achieve an 80% FTE, is difficult. To achieve "success" in academics can require a commitment of 70–80 hours/week. A faculty member is unlikely to be willing to work a 60 h "part-time" job. Many responsibilities, meetings and clinical activities do not provide the flexibility needed to meet a part-time commitment there by slowing an individual's professional advancement more than a 20% decrease in hours would imply [20]. The implementation and success of a part-time strategy for faculty must be achieved with realization of numerous complicating issues including perceptions by senior faculty of a lack of commitment, a system for fair distribution of clinical responsibilities and incentives, and maintaining clinical and educational skills.

2.10 Summary

The combination of generational and gender shifts in the academic physician population has dramatic implications for work force predictions particularly in academics. No longer can it be assumed that a new graduate will fill the role of the retiring physician. The new generation of physicians is generally not working the long hours of the baby boomer generation and may not see as many patients in a given week. Some see this as a generation that is less dedicated to their patients and not as hard working. This is not true. They are dedicated to their patients and specialty. However, they also hold their families and personal health as a high priority. Their identities are based on more than their profession. Both men and women look to practice settings that give them the flexibility of professional growth and satisfaction as well as personal growth. Women may want to work part time when children are small and then move to a more full time status as they enter school. Fathers want more an active roles to coaching teams, having actual family vacations and attending parent-teacher conferences. Without a major influx of new physicians, which is not going to happen for a number of years given the 8–12 year medical educational pipeline, health care delivery systems must develop creative ways to accommodate a variety of practice styles into an efficient, cost effective, professionally satisfying, team building approach to care. The private sector is commonly seen as more flexible and creative with part time jobs and roles such as laborist. To remain competitive in the workforce for the best and the brightest, academic departments must become more creative and open to variations in work settings and expectations for junior faculty to be able to recruit and retain the individuals who will train the next generation and develop the next wave of medical/surgical innovations.

References

1. Schaffner, H.A., and C.E. Van Horn. 2003. *A nation at work*. New Brunswick, NJ: Rutgers University Press.
2. Staiger, D.O, I.D. Auerbach, and I.P. Buerhaus. 2010. Trends in the work hours of physicians in the US. *JAMA* 303:747–753.
3. Kotlikoff, L.J., and S. Burns. 2005. *The coming generational storm*. Cambridge, MA: The MIT Press.
4. U.S. Department of Education, National Center for Education Statistics, Digest of Education Statistics, 2006.
5. Moody, J. 2002. Recruiting generation X physicians. *The New England Journal of Medicine*, Jan–Feb 2002.
6. Bickel, J., and J.A. Brown. 2005. Generation X: Implications for faulty recruitment and development in academic health centers. *Academic Medicine* 80(3):205–210.
7. Berthold, J. 2008. Is the generation gap a growth opportunity? *American College of Physicians*.
8. Keeter, S., and P. Taylor. 2009. The millennials. Pew Research Center. http://pewresearch.org/pubs/1437/millennials-profile. Accessed 11 Dec 2009.
9. Kirch, D.G. 2007. From Marcus Welby to Grey's anatomy: The next generation. *Medscape*.

10. Howe, N., and W. Strauss. 2000. *Millennials rising, the next generation*. New York, NY: Vintage Books.
11. Center for Workforce Studies. 2008. Physician Specialty Data, AAMC 2008.
12. Zare, S.M, A.J. Galanko, E.K. Behrns et al. 2005. Psychologica well being of surgery residents after inception of the 80-hour workweek: A multi-institutional study. *Surgery* 138:150–157.
13. Manion, J. 2005. Creating a positive health care workplace! practical strategies to retain today's workforce and find tomorrow's, health forum, Inc. *As noted In*: Reduce physician turnover and improve your bottom line – Physicians speak up about retention issues. www.LocumTenens.com/retention.
14. Clarke, M.C, C.L. Koch, and J.E. Hill. 2004. The work-family interface and differentiating balance and fit. *Family and Consumer Sciences Research Journal* 33(2):121–140.
15. Autry, A., D. Irby, and C. Hodgson. 2007. Faculty attrition in obstetrics and gynecology. *AJOG* 196:603–608.
16. Brotherton, S.E., P.H. Rockey, and S.I. Etzel. 2004. Graduate medical education. *JAMA* 292(9):1035.
17. Rayburn, W.F, L.B. Anderson, V.J. Johnson et al. 2010. Trends in the academic workforce of obstetrics and gynecology. *Obstetrics and Gynecology* 115:1–6.
18. Lieff, S.J. 2009. The missing link in academic career planning and development: Pursuit of meaningful and aligned work. *Academic Medicine* 84(10):1383–1388.
19. Bunton, S.A. 2008. Differences in US medical school faculty job satisfaction by gender. Analysis in brief. *AAMC* 8(7).
20. Helitzer, D. 2009. Missing the elephant in my office: Recommendations for part-time careers in academic medicine. *Academic Medicine* 84(10):1330–1331.

Chapter 3
Clinician Educators: How Can We Meet the Expanding Need?

Joanna M. Cain and Kathleen C. Bowling

> *The American medical school is now well along in the second century of its history. It began, and for many years continued to exist, as a supplement to the apprenticeship system still in vogue during the seventeenth and eighteenth centuries. The likely youth of that period, destined to a medical career, was at an early age indentured to some reputable practitioner, to whom his service was successively menial, pharmaceutical and professional: he ran his master's errands, washed the bottles, mixed the drugs, spread the plasters, and finally, as the stipulated term drew towards its close, actually took part in the daily practice of his preceptor...*
>
> Flexner [1]

Abstract Distinctions are made between faculty who are physician scientists and clinician educators in women's health care. As departments expand due to increasing clinical and educational needs, the clinician educator will play a more central role. This chapter focuses on the expanding value and need for clinician educators.

Keywords Clinician educator · Ambulatory education · Faculty motivation and reward

3.1 Introduction

Medical education after Flexner concentrated on bringing the theoretical and scientific basis of medicine into the center of medical education as well as standardizing medical education across medical schools. Flexner's ideal would have been a clinician researcher as the primary educator who could seamlessly go from the bench to

J.M. Cain (✉)
Department of Obstetrics and Gynecology, Warren Alpert School of Medicine of Brown University, Providence, RI, USA
e-mail: joanna.cain3@gmail.com

W.F. Rayburn, J. Schulkin (eds.), *Changing Landscape of Academic Women's Health Care in the United States*, International Library of Ethics, Law, and the New Medicine 48,
DOI 10.1007/978-94-007-0931-7_3,

the bedside, imparting that evidence based knowledge skillfully and even subjecting the process of education to the research scrutiny that would advance techniques and outcomes. Clinician Educators are the key to robust women's health education – particularly community and ambulatory based education. Yet now more than 100 years from Flexner, we have yet to find the right balance of Boyer's [2] scholarships of discovery, integration, application and education either in our educational process or in our development as an academy of professors of medicine that is required to promote education in sex and gender based medicine. Clinician educators have generally failed to attain status as part of the "first class" professoriate in many medical schools [3], and the status of schools is often measured by the level of NIH funding rather than the quality of education – partially because adequate metrics to assess scholarship remain less developed for clinician educators. Research to develop these metrics for clinical competency and humanism has little to no funding. The Flexner report itself has been criticized as widening this gap between biomedical science and clinical medicine and education [4].

The bulk of the educational process of clinical learning is based in various rotations in academic and non academic hospitals and ambulatory clinics staffed by clinical faculty who learned from those who came before them. This education includes both the application of this knowledge to patients and the application of theories of professionalism and education that are often unique to a given institution. Some rotations in the time since Flexner have been highly professional with long standing and highly revered teachers who thought about the actual process of education and taught the best of medicine, of bedside manners and professional conduct. Others were less salutatory sites of learning with bursts of anger and student abuse being normative, intentional malice between professionals, and patients seen only as a means to an end. Socialization into medicine and learning in the clinical setting remained variable.

Combining the scholarships of discovery, integration, application and education (teaching) into coherent clinical education is a challenge that we have yet to fully complete. For women's health, for which both focused research and educational scholarship is still developing, the gap may be even wider.

Historically, the construct and management of schools of medicine were shaped by faculty who valued a paradigm of the primacy of the discovery/integration scholarships and generally did not value nor bear the responsibility for application and educational scholarships. The scholarship of discovery (bench and to some degree translational research) still holds primacy over the others in most schools. Most promotion committees do not have metrics to fully evaluate the scholarship of education and/or they do not consider this scholarship of equal value. This leads to lack of broad exploration of the scholarship of education and lack of a robust faculty with both generalist and specialist scholars in teaching and education in virtually every discipline. In particular, the continued gap in advancement of women in academic medicine can be tracked to lack of valuing the scholarship of education equally in most settings.

We are now in the era in which variation in educational scholarship and in professionalism is no longer acceptable. The setting in which clinical education is carried out is often under very limited oversight of Departments or Schools of Medicine, and can have conflicting values or interests (financial, program competition, faculty competition, etc.) that have to be adjudicated to even begin to focus on the educational scholarship. The ability to shape the capacity to teach and develop scholarly educators is hampered to varying degrees by lack of time, lack of interest, and lack of meaningful rewards or consequences for providing or NOT providing the highest levels of teaching or professionalism. Larger and larger faculties have been hired to provide the clinical educational activities. They have become more important to schools of medicine as a means of broadening the institution's economic base – often directly in conflict with the time that is needed to develop as a clinician educator. Furthermore, those areas that are not highly compensated, for example many aspects of women's health within Ob/Gyn and General Internal Medicine, are often under more financial pressure and have less ability to subsidize educational or research development. Similarly, the increased value given to the scholarship of discovery in schools of medicine can result in a conflict of interest to keep star researchers, who may not be held equally accountable for poor teaching skills and poor professionalism. The end result is that most of medical education looks to some level of a "voluntary" or community based faculty to augment some or even provide

Table 3.1 Roles of clinician educators

Timeline	*Role*
Pre Flexner	*Oversees apprentice system: primary focus of medical education is entirely clinician educators with very little research or educational scholarship expected*
Post Flexner: initial	*Initial division of basic knowledge to University setting, with clinical rotations still heavily dependent on community clinical rotations. Significant growth of scholarship of discovery (research) and development of scholarship of education.*
Contemporary	*Curricular harmonization across United States; growth of medical school faculties with cross subsidy of education and research from clinical enterprise. More of education or all of education within University clinical enterprise but still heavily dependent on "voluntary" or community clinical educators*
Developing trends: increased student and resident numbers; clinical enterprise dependence of SOM with variable viability and lacks capacity for covering incoming students; faculty retention for clinical educators challenged by promotion and tenure policies and salaries; community educators challenged by decreasing reimbursement and increasing time in practice leaving little time to teach.	***Growth in dependence on community based clinical educators likely, continued development of balanced educational, scholarship, recognition, and promotion strategies for clinician educators will be needed.***

the bulk of experience for students and residents. These faculty members have varying degrees of commitment to their trainees and their success and hence, may result in the varying outcomes that the Flexner report critiqued so clearly a century ago (Table 3.1).

This chapter seeks to explore the role and future development of clinical educators. What motivates them to teach? What heights of competency in educational scholarship and professionalism could be achieved while confronting the shifting grounds of medical funding and stress that affects our faculty – both institutionally and community based? What could uniquely be done in women's health to expand and develop these roles?

3.2 What Motivates Clinician Educators?

There is an element in medicine that is often overshadowed by all the contortions inherent in the business of medicine, malpractice issues, and generational differences – the simple desire of virtually all individuals who practice medicine to leave a legacy of making a difference in their patients' lives and in the practice of the next generation of physicians. As with all of medicine, clinician educators experience the "paradoxical presence of self interest and altruism in the tradition of Western medicine. . .the responsibility that uses skills well for self-satisfaction and self-support; in this, medicine adheres to the principle of self-interest. It will, at the same time, be the responsibility that uses skills to respond to the needs of others; in this, medicine is ruled by the principle of altruism. This moral paradox, like a geographic fault, penetrates the terrain of medicine" [5].

This moral paradox penetrates the academy of clinician educators as well, particularly as we search for the next generation of educators from a pool of graduates who have a higher financial burden than ever before and no clear funding or its more important correlate, time, for scholarly development as clinician educators (Fig. 3.1). Both monetary and altruistic motivations have to be considered. Asking for them to teach in their clinical setting clearly adds financial burdens by increasing the time it takes for them to see a patient for most medical student level preceptors, although resident preceptors often gain efficiency at the senior levels – a recognized trade off in residency education [6, 7].

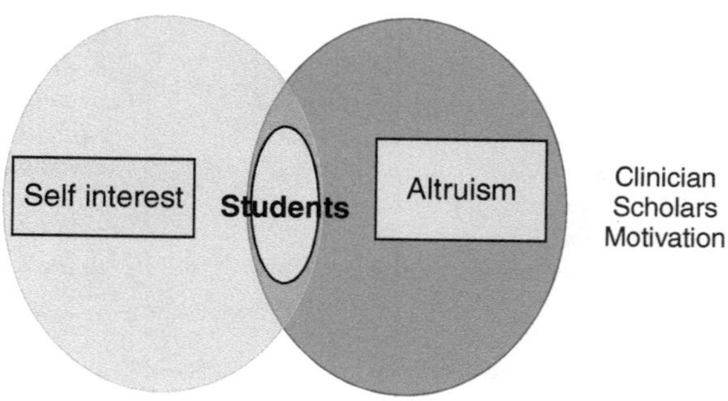

Fig. 3.1 Teaching crosses both the motivation of self interest and altruism for clinician educators

Providing a stipend as a motivation for community and hospital based educators would then seem to be a motivational solution to increasing the number of clinician educators. The provision of a nominal stipend has been explored in the primary care setting [8]. While it directly impacted retention of teachers, it was not the major source of satisfaction – echoing the moral paradox of medicine outlined by Jonsen. The majority ranked having a good student as the major factor in continuing to teach (73%) as well as satisfaction, while only (8%) rated the stipend as the most important factor and none rated it as important in satisfaction. The stipend was seen as both gap coverage for the fewer patients seen during precepting sessions as well as recognition of the value of their teaching. Paying physician teachers for their efforts has also been shown to correlate with improved student experience evaluations [9]. In acting as clinical preceptors and longitudinal advisors, again minimal support resulted in gains across all areas (overall, critique of student presentations, and critique of student write ups, establishing rapport, physical exam, and history taking) in student satisfaction.

Monetary rewards are not the only motivators for clinician educators (Table 3.2). Volunteer, community based faculty have been shown in many studies to provide precepting for the simple satisfaction gained from teaching, the opportunity to be challenged and the prospect of influencing others by serving as role models [10, 11]. They also have been shown to value student feedback [12], likely as part of the satisfaction of being challenged and teaching. Non-monetary rewards, such faculty appointments, teaching awards and educational materials and continuing medical education credit also have significant value for community faculty [10–12].

The intrinsic motivations of clinician educators and preceptors remains very altruistic – demonstrating what community practice is like, enjoyment of teaching, giving something back to their profession, intellectual stimulation and being a role model. In women's health, often the chance to demonstrate gender or sex specific medical issues to expand a student perspective has particular attraction. Clinician educators want to leave a legacy that they made a difference that will live on in their students and mentees – the immortality of all great educators. Reward and recognition is valued, and does encourage retention and expanded efforts as well as student satisfaction that assures continued motivation to teach and precept – but it is important to remember that within the paradox of medicine, the primary motivation for teaching stems from altruism not self interest.

Table 3.2 What motivates clinician educators? Both intrinsic and external motivation is involved

Intrinsic motivation	External motivation
Sense of legacy and continuity	Recognition from student evaluations/feedback
Stimulation in science, clinical care, and development as an educator	Monetary stipend
Fellowship/friendship with other educators	Faculty appointments, teaching awards
Pride in medicine ("having a good student", the enthusiasm of students)	CME credits

3.3 How Do We Create Excellence and Scholarship in Clinical Educators?

There is no doubt that early and continuing exposure to medical education in clinical settings molds students – both by improving socialization into the profession and influencing what disciplines or sub disciplines they consider for themselves [13]. In undergraduate settings, students note the significant learning they gain from ambulatory settings by witnessing the impact of long term patient relationships, the impact of social environment on disease processes, and the importance of dealing with people not diseases [14, 15]. In the hospital setting, learners and teachers valued hands on experience, broad knowledge transfer and working in small groups with clinicians [16]. What is most remarkable about education in the clinical setting is that the majority of clinician educators are recruited to teach without ever having any formal training or support for the skill sets and the underlying pedagogy that could make their efforts more successful. Often the only guide is the way in which they were taught – for good or for ill.

It is clear that while many clinical teachers adopt the methods they were taught with, they are hungry to explore improvements in methods. One study in Australia [16] noted that facilitation of sessions fostering high level cognitive processing from pre reading before clinical experiences was a skill set that required development of more advanced teaching strategies and skills than the clinician educators presently possessed. In addition, professional development with new technologies such as software and online learning, simulation, and development of mentoring skills have all been identified as areas for clinician educators to master.

Studies of programs to develop new skill sets in clinician educators are growing. A systematic review of faculty development initiatives concluded that overall satisfaction with faculty development programs was high and resulted in positive change in attitude, knowledge, and teaching behavior [17]. This review includes a summary of the eight most highly rated (design, outcomes) studies of faculty development. There were some common findings particularly that baseline performance improvement is possible, and that those with the worst evaluations stand to benefit most. In addition the reviewed studies suggest that individual variation in skill sets requires tailoring interventions to accommodate different needs. They also note the importance of tailored and persistent feedback without which teaching problems were likely to persist. The interventions discussed include videotaping with intensive feed back, multiple 2 h seminars, and full fellowships as the vehicles for learning. Those of a more longitudinal nature were more likely to result in longer term gains and development of social networks of educators that enhanced the changes. An additional key finding was the importance of peers as role models, to facilitate exchange of ideas and the critical nature of collegial support to promote and maintain learning and change.

The importance of peers in faculty development was highlighted in a program at University of California, Los Angeles. The program focused on teaching skill development in the setting of small groups of community faculty led by a local opinion leader [18]. This model used a "Teach the Leaders" group initially (with dinner and a small stipend leading to 80% attendance) to develop a case based curriculum. Feedback and teaching tools (videos, readings, tutor guides) were provided to this leadership group with the plan that they would teach their peers in the community (again in small groups). The cases they taught from had roles for students as well as teachers and were positively received by students as well as community faculty. There is little evidence based literature on gender differences in faculty learning as teachers in medicine. However given the predominance of women in medicine as faculty for women's health clerkships, this is an area that deserves further exploration.

Educational fellowship programs can be useful in developing the skills of educational scholars and peer leaders [19]. The goals of these long term programs were to prepare educational scholars to "lead, design, implement, and evaluate educational initiatives" and went far beyond the usual focused faculty development in teaching skills. It is notable that the success of these programs has been related to continuous feedback and adaptation to changing needs of the professoriate/community faculty and to changing priorities.

One other characteristic in development of these scholar programs is important as we consider how to sustain clinician educators. There is a critical need to create cross disciplinary relationships and scholarship to provide a large enough community for robust development of education and educational research through the program design. Virtually all the courses have incorporated that into their design. This is particularly important in women's health where the critical skill sets for both undergraduate and graduate programs are acquired only through cross disciplinary training, requiring cross disciplinary competencies in curriculum design and instructional techniques [20]. In the Teaching Scholars Program at McGill University, for example, 15 different specialties have so far been chosen as Scholars [21]. The "community of practice" that has developed among the scholars is one of the most positive outcomes from the scholar's evaluations.

The McGill program provided University level courses from the Education Faculty, a monthly seminar, educational project requirement as well as the faculty development activities. The University courses allow health professionals entrée into the world of pedagogy of education and gives them new frameworks and ideas for their own courses and research. At UC Davis, the goal of the Teaching Scholars program was to "enhance faculty understanding of medical education's complexity, and improve educators' effectiveness as regional/national leaders [22]." This was structured with a 1/2 day/week over 24 weeks and incorporated the key idea that creating change required reflection on present practice. The four foundational principles of their program are pertinent to all approaches to educating the educators:

1. *Theory relevant to medical education powerfully anchors reflection-on-practice.*
2. *Diversity of perspectives on educational practice can promote reflection*
3. *Colleagues and peers are often the best teachers for each other*
4. *Reflection is a reciprocal process of testing theory in practice and examining practice in the context of theory.*

They used an institution wide design process focused on recognizing and addressing potential barriers. As an example, one of the strategies for the scholars was to address lack of overt reimbursement for faculty teaching with better evaluation of teaching expertise and leadership to improve evaluations needed for promotion. The scholars noted that they diversified evaluation methods of learners' skills, were better at systematic planning and more confident in skills and much more able to lead participative teaching, problem solving and take on new projects.

These findings are echoed in the program outcomes at the University of Washington (UW) [23] and Beth Israel Deaconess/Harvard Medical School (BIDH) [24]. The challenges cited by the programs include the very stress of time demands for scholarship, clinical care, lack of funding, and educational endeavors that challenge all of clinical education in medicine. Funding in particular remains a central challenge to all the fellowship programs. The time demand element is increasingly important as faculty choose part time work [25] with the lost revenue to fund time for initiatives such as a faculty scholarship in education. Given the likelihood of young faculty choosing the part time route and the fact that this might well be the exact time that such an educational scholar program is most effective in their career – funding and time will continue to be challenges that must be met if we are to have the educational leadership we need for medicine.

Beyond the expert educator programs, there have been attempts to outline those key pedagogical concepts that are the most important to help clinical teachers better understand their work and improve. A group of international experts came up with 4 broad categories that can form the basis of any faculty development from minimal to scholar programs [26]. These included: Curriculum, Adult learning, Helping adults learn, Assessment as organizing areas for key concepts such as summative and formative assessment: goals, objectives, learner differences, motivation to learn. Key areas for clinical faculty baseline and advanced development drawn from these programs are outlined in Table 3.3.

It is clear that assuring we have excellence in clinical education requires more than simply telling individuals to teach, and assuming that simply the knowledge and skill set they have acquired with their own training and experience will lead to

Table 3.3 Opportunities for clinical faculty development programs

All faculty development	Advanced development
Review and critique of curricula, goals, and objectives for their rotation/mentorship	Development of Curriculum
Small group teaching and mentorship	Competency based education/evaluation
Basic theories of adult learning pertinent to their setting (including feedback)	Learning theory and development of clinical reasoning skills
Teaching procedural skills/competency based learning	Simulation learning, theory, practice
Technology to improve teaching efficacy	Development of new technology and materials for teaching

an effective form of education. Teaching scholar programs and faculty development programs all show improved efficacy and satisfaction for the faculty.

3.4 Recognizing the Value of Clinician Educators and Community Faculty

Given the reliance of every medical school on clinical faculty for more than half of the educational development of medical students and significantly more for fellows and residents, it is surprising that we have not developed robust and sensitive criteria for promotion and recognition on the basis of educational contribution and scholarship within the professoriate of medical schools [27]. For women's health with its dependence on the least compensated members of the professoriate disciplines, recognizing and supporting their value may become critical in the challenging economics we face. Clinician educators often do not have the time or the infrastructure support within academic medicine to publish or promote their innovations in education beyond the institution. Unlike research development and infrastructure, there is virtually no infrastructure extant for assuring national and international educational materials development, pilot funding, and promotion at institutions. They also, as we have noted in the prior section, rarely have the training that would allow them to use the jargon and concepts of the pedagogy of education to develop publishable studies and work. The markers for their work are often at the best teacher, mentor or role model level – but that alone is insufficient for recognition in the usual promotion process. Even separate tracks for clinician educators fail to recognize this inherent catch 22 in the structure they are faced with – regional and national reputations as well as the lack of reliable measures for success hamper their success. Some authors [28] even suggest that holding clinician educators to the criteria for regional and national reputations might hinder their efficacy as teaching scholars, given the extraordinary time demands of creating excellence as a teacher and educational scholarship in their own institution. Failure to address this is likely to create high turnover and less commitment to teach and precept in the institution based and community based faculties, as well as lead to morale issues and loss of sustained gains in educational development and scholarship. The same authors imagine an RO1 mechanism for educational research, a broad infrastructure to nurture these faculty, and recognition through promotion. Scholarly work can be defined as meeting the requirements that it be open for critical review and exchange and use by other scholars (egg. It is public) which begins to create a platform on which more coherent promotion and recognition strategies can be built [29, 2].

Separating tracks into a "clinical" versus "academic" may allow for clarity around the what regional or national recognition is appropriate for the clinician educator and provide clarified expectations and recognition for often significant innovation and body of work by faculty for whom the primary focus is clinical work, whether community or institution based. The rigor of clinical versus academic track expectations, however, needs careful design to avoid detrimental diminishing of the value of clinical track designation.

Tools to analyze the extent to which faculty have achieved scholarly work within the scholarship of education may also facilitate recognition [30]. Many institutions have begun educational portfolios, although some are too elaborate or designed from non medical settings where arenas such as surgical teaching and simulation do not fit into the templates easily. A simple excel file [31] may enhance this activity and also give a template for teaching awards that can additionally assist in promotion activities. These can assist in creating an infrastructure and analysis more suited for educational scholars rather than the research scholars more traditionally evaluated by promotion and tenure committees.

In the prior discussion about motivation, the reward of a stipend was considered for expert teachers/leaders and shown to be valued by the faculty. The authors [8] concluded that there was a level beyond a mere token of thanks that allowed busy practitioners to focus on the pleasure of teaching. Given the tight financial straits of most institutions – the possibility of stipends to recognize and value clinician educators is usually limited, but applying it to indicate the value of the educational enterprise for at least the leaders of that enterprise can have a positive effect.

Teaching awards and recognition of peers remain central elements that value our clinician educators and continue to be encouraged and expanded. However, without broad recognition of their value within the medical professoriate and academic medicine, and without the infrastructure and support for development of educators and the educational enterprise, we will continue to risk burn out of those who provide the backbone of medical education and fail to achieve the educational experience that produces the broadest skill set and competencies for the future health of the world and of women.

3.5 Steps to a New Future: Confronting the Ivory Tower – Does the Future Lie in Increasing or Decreasing Medical School Faculties in Favor of Clinical Site and/or Community Faculty?

We do not actually know the true costs of medical education at the undergraduate or graduate levels or the variation between hospital settings and the ambulatory setting [32]. Faculties of medicine have increased astronomically to meet the burgeoning need for clinical revenue without concurrent development of the infrastructure to enhance clinical faculty development as educators or to develop and fund the educational research needed or advance the innovative educational programs developed. The imbalance between clinical and research emphasis led to a redefinition of scholarship "determined by research publications and often measured in terms of quantity, since quality was difficult to evaluate. Highly knowledgeable clinicians, with unquestioned excellence in erudition, judgment, patient care, and teaching, were often denied tenure and lost from academic faculty because of deficient scholarship. They had not published enough" [33].

Concomitant with the focus on publication and research scientists as the luminaries of medicine, schools lost the natural alignment with community faculty borne of valuing their contributions – particularly those of patient care and teaching. At the same time, schools of medicine have come to depend even more heavily on community faculty who serve out of the motivations of legacy and giving back and usually without additional support or recognition other than the occasional teaching award. Creation of clerkship rotations depends on these motivations and the charisma of various clerkship leaders to engage institutional and community faculty in this endeavor. At present, we are limping along but a shadow of what can and should be achieved for medical education and scholarship. We have not fully wrestled with what it will take for individual learners to feel fully confident within their discipline to provide the highest quality of care for women, nor have we fully developed a faculty with the skills to assure that those individual needs and learning styles can be accommodated. The contemporary faculty is facing a tipping point at which the increasing financial stress on community faculty particularly those focused on women's health, the lack of recognition of educational scholarship and excellence in promotion processes, and the lack of funding will create a true crisis for medical education at all levels in the United States. It is also a point of opportunity to truly rebalance "the values of humanism and science and the development of innovative curricula to support the full power and scope of care that physicians deliver to their patients" [34] and achieve the integration that Flexner left as a dilemma between biomedical science and clinical science [4].

There are positive steps to develop and support clinical educators that all institutions can take now. We must first support and celebrate the altruistic motivation of faculty with meaningful awards and recognition of the ways in which they make a difference with education and the legacy that they leave. Secondly, develop whatever level of infrastructure support is feasible economically as soon as possible, including cross disciplinary faculty development, educational scholars programs and academies, peer support groups, grant and funding mechanisms and support for development of educational intellectual property within the institutional intellectual property format. Third, advocate for robust and perhaps multiple avenues for clinician educator promotion that derive their promotion criteria from relevant innovation and scholarly educational endeavors and achievements – including advancing women's health education throughout the undergraduate and graduate medical curriculum.

References

1. Flexner, A. 1910. *Medical education in the United States and Canada*. The Carnegie foundation. Boston, MA: Merry Mount Press, 3 (Digitalized by Google).
2. Boyer, C. 1990. *Scholarship reconsidered: Priorities of the professoriate*. Princeton, NJ: The Carnegie Foundation.
3. Cooke, M. et al. 2006. American medical education 100 years after the Flexner report. *NEJM* 355:1339–1344.

4. Boudreau, J.D., and E.J. Cassell. 2010. Abraham Flexner's "mooted question" and the story of integration. *Academic Medicine* 85:378–383.
5. Jonsen, A. 1990. *The new medicine and the old ethics*. Cambridge, MA: Harvard University Press, 7.
6. Bowen, J.L., and D.M. Irby. 2002. Assessing quality and costs of education in the ambulatory setting: A review of the literature. *Academic Medicine* 77:621–680.
7. Latessa, R. et al. 2007. The satisfaction, motivation, and future of community preceptors: The North Carolina experience. *Academic Medicine* 82:698–703.
8. Peters, A.S. et al. 2009. How important is money as a reward for teaching? *Academic Medicine* 84:42–46.
9. Ashar, B. et al. 2007. An association between paying physician-teachers for their teaching efforts and an improved educational experience for learners. *Journal of General Internal Medicine* 22:1393–1397.
10. Fulkerson, P.K., and R. Wang-Cheng. 1997. Community based faculty: Motivation and rewards. *Family Medicine* 29:105–107.
11. Kumar, A., D.J. Kallen, and T. Mathew. 2002. Volunteer faculty: What rewards or incentives do they prefer? *Teaching and Learning in Medicine* 14:119–123.
12. Dent, M.M., J. Boltri, and I.S. Okosun. 2004. Do volunteer community based preceptors value students' feedback. *Academic Medicine* 70:1103–1107.
13. Dornan, T. et al. 2006. How can experience in clinical and community settings contribute to early medical education? A BEME systematic review. *Medical Teacher* 28:3–18.
14. Howe, A. 2001. Pateint-centered medicine through student-centered teaching: A student perspective on the key impacts of community based learning in undergraduate medical education. *Medical Education* 35:666–672.
15. Major, S.C., and P. Booton. 2008. Involvement of general practice in undergraduate medical education in the United Kingdom. *Journal of Ambulatory Care Management* 31:269–275.
16. Young, L. et al. 2009. Effective teaching and learning on the wards: Easier said than done? *Medical Education* 43:808–817.
17. Steinert, Y. et al. 2006. A systematic review of faculty development initiatives designed to improve teaching effectiveness in medical education: BEME guide no.8. *Medical Teacher* 28:497–526.
18. Wilkes, M.S. et al. 2006. An innovative program to augment community preceptor's practice and teaching skills. *Academic Medicine* 81:332–341.
19. Gruppen, L.D. et al. 2006. Educational fellowship programs: Common themes and overarching issues. *Academic Medicine* 81:990–994.
20. Cain, J.M. et al. 2002. Why is it so important to ensure that women's health, gender based competencies are woven into the fabric of undergraduate medical education? *American Journal of Obstetrics and Gynecology* 287:S1–S3.
21. Steinert, Y., and P.J. McLeod. 2006. From novice to informed educator: The teaching scholars program for educators in the health sciences. *Academic Medicine* 81:969–974.
22. Srinivasan, M. et al. 2007. Developing the master educator: Cross disciplinary teaching scholars program for human and veterinary medical faculty. *Academic Psychiatry* 31:452–464.
23. Robins, L., D. Ambrozy, and L.E. Pinsky. 2006. Promoting academic excellence through leadership development at the University of Washington: The teaching scholars program. *Academic Medicine* 81:979–983.
24. Hatem, C.J., B.A. Lown, and L.R. Newman. 2006. The academic health center coming of age: Helping faculty become better teachers and agents of educational change. *Academic Medicine* 81:941–944.
25. Harrison, R.A., and J.L. Gregg. 2009. A time for change: An exploration of attitudes toward part time work in academia among women internists and their division chiefs. *Academic Medicine* 84:80–86.
26. McLeod, P.J., Y. Steinert, T. Meagher, and A. McLeod. 2003. The ABCs of pedagogy for clinical teachers. *Medical Education* 37:638–644.

27. Levison, W., and A. Rubenstein. 2000. Integrating clinician-educators into academic medical centers: Challenges and potential solutions. *Academic Medicine* 75:906–912.
28. Levison, W., and A. Rubenstein. 1999. A mission critical: Integrating clinician educators into academic medical centers. *New England Journal of Medicine* 341:840–844.
29. Collins, J. 2004. Teacher or educational scholar? They aren't the same. *Journal of the American College of Radiology* 1:135–139.
30. Chandran, L. et al. 2009. Evaluating the performance of medical educators: A novel analysis tool to demonstrate the quality and impact of educational activities. *Academic Medicine* 84:58–66.
31. Schindler, N., D.P. Winchester, and H. Sherman. 2002. Recognizing clinical faculty's contributions in education. *Academic Medicine* 77:940–941.
32. Bowen, J.L., and D.M. Irby. 2002. Assessing quality and costs of education in the ambulatory setting: A review of the literature. *Academic Medicine* 77:621–680.
33. Feinstein, A.R. 2003. Scholars, investigators, and entrepreneurs: The metamorphosis of American medicine. *Perspectives in Biology and Medicine* 46:234–53.
34. Rabow, M.W. et al. 2010. Professional formation: Extending medicine's lineage of service into the next century. *Academic Medicine* 85:310–317.

Chapter 4
Interdisciplinary Women's Health Research and Career Development

Vivian W. Pinn and Mary C. Blehar

Abstract Since the 1970s, many leaders in biomedical research have warned about a looming national shortage of new physician investigators but evidence, especially for Women's Health, has been largely indirect or anecdotal. This chapter discusses the number of junior physician investigators, in either patient-oriented or basic science research, who are present in departments and who are needed to maintain a research mission.

Keywords Biomedical research careers · Interdisciplinary research · Physician scientists · Women's health research · Sex differences research

4.1 Historical Introduction

With the advent of Federal programs to enhance and expand knowledge about the health of women, the concept of what constitutes the field of women's health research has evolved and expanded in scope. These changes in turn have had a major influence on Federal priorities for women's health research and for the career development of those who will study women's health, deliver their care, and determine changes in public policy.

During most of the 20th century, the traditional interpretation of women's health was almost synonymous with reproductive health, and women's health research primarily addressed the reproductive system. This limited focus excluded a wide range of other serious conditions and diseases, especially those affecting both sexes. In the 1970s and 1980s, with the rise of feminism in the US., women's health advocates and scientists alike began to voice concerns about the lack of inclusion of women in clinical studies of conditions that affect both men and women. They decried what they called a "bikini approach" to women's health, focused on the breast and reproductive system, while the rest of the woman was ignored [1–4]. Increasingly

V.W. Pinn (✉)
Office of Research on Women's Health, Department of Health and Human Services,
National Institutes of Health, Bethesda, MD 20817-5484, USA
e-mail: vivian.pinn@nih.gov

W.F. Rayburn, J. Schulkin (eds.), *Changing Landscape of Academic Women's Health Care in the United States*, International Library of Ethics, Law, and the New Medicine 48,
DOI 10.1007/978-94-007-0931-7_4,

they voiced criticisms of the biomedical community for its acceptance of a "male model" of health and disease, which was predicated on a largely unexamined belief that women and men reacted comparably to diseases and to the interventions to prevent and treat them. This perspective ignored the role that sex and gender differences may have in the incidence, prevalence and course of a number of major chronic diseases [5, 6].

The absence or relative dearth of women as participants in clinical trials of conditions not unique to women perpetuated the male model as the norm [7–9]. Examples of exclusion of women in some early major clinical trials for the prevention of cardiovascular disease–acknowledged as the leading cause of premature mortality in women–were particularly troublesome. One such example was the Multiple Risk Factor Intervention Trial (the MRFIT study), a large-scale National Institutes of Health (NIH) funded clinical trial, which demonstrated the value of preventive interventions to reduce health risks from smoking, high blood pressure and elevated cholesterol level. A primary endpoint, death from coronary disease, was reduced by the interventions. However, the study recruited only male subjects [10, 11]. The findings another large-scale NIH-funded study, the Physicians Health Study, which demonstrated the efficacy of aspirin in the primary prevention of cardiovascular disease, were also based on an entirely male sample [12]. Findings of a 2005 study that tested the preventive benefits of aspirin in women challenged the presumed lack of sex differences in effect; [13] this later study indicated that aspirin reduced women's risk of stroke but not of cardiovascular disease.

In 1985, a Public Health Service (PHS) Task Force on Women's Health was convened to address gaps in knowledge about women's health through Federal policy [14]. The Task Force's report was highly influential in leading NIH in 1987 to promulgate guidelines for the research community that *urged* the inclusion of women in NIH-funded research [15]. (Later that year, NIH recognized the need to address the inclusion of minority populations and a revised 1987 version of the NIH guidelines *encouraged* the inclusion of minorities in clinical studies.) Not satisfied that the NIH inclusion guidelines were achieving their desired results, in 1990 the *Congressional Caucus for Women's Issues* asked the US. General Accounting Office (GAO) to conduct an investigation into their implementation. The resulting report criticized implementation, noting that it was slow and uneven among NIH Institutes and Centers (ICs). The report noted that the guidelines had not been well-communicated to NIH staff and the scientific community and that gender analysis was not being carried out routinely, so that the scientific impact of the inclusion guidelines could not be fully realized. The GAO report further indicated that not all ICs factored adherence to the inclusion guidelines into their scientific merit review [16].

In 1990 the NIH Acting Director, Dr. William Raub, responded to Congressional concerns about the lack of consistency in implementation of the inclusion policy by establishing the Office of Research on Women's Health (ORWH) within the Office of the NIH Director and under the initial acting leadership of Dr. Ruth Kirschstein [17] who was co-chair of the PHS Coordinating Committee on Women's Health Issues at that time. The Office was charged with ensuring that NIH-supported research appropriately addresses women's health issues and includes women in clinical research, especially in clinical trials. ORWH was directed to establish NIH-wide women's health research goals and policies and to coordinate NIH activities undertaken in performing such research. The Office was designed to serve as the NIH point of contact on women's health research for external scientific and medical communities, other organizations interested in women's health, and other components of government.

With the establishment of ORWH, NIH began earnest efforts to develop a research agenda based on advocacy, scientific, and professional considerations that would address gaps in scientific knowledge concerning women's health and would increase the number of scientists undertaking such research. These aims were articulated in the first ORWH agenda setting report, *Opportunities for Research on Women's Health* [18]. NIH also began to reinforce efforts for the inclusion of women in clinical research.

In order to ensure that the policies for inclusion were fully implemented by NIH, Congress made what had previously been policy into Public Law through a section in the NIH Revitalization Act of 1993 entitled *Women and Minorities as Subjects in Clinical Research* [19]. This same Act also established ORWH in statutory language, including provisions for the Office to ensure compliance with the language related to inclusion. The Revitalization Act essentially reinforced existing NIH policies, but added four new requirements: (1) that women and minorities and their subpopulations are included in all NIH clinical research; (2) that women and minorities and their subpopulations are included in Phase III clinical trials in numbers adequate to allow for valid analyses of differences in intervention effect; (3) that cost is not allowed as an acceptable reason for excluding women and minorities from NIH research; and (4) that NIH initiate programs and support for outreach efforts to recruit and retain women and minorities and their subpopulations as participants in clinical studies. Revised NIH inclusion guidelines developed in response to this law were published in the *Federal Register* in March 1994 and became effective in September 1994 [20]. They articulated a new NIH policy that any extramural grant, cooperative agreement or contract or any intramural project that did not comply with the inclusion policies would not receive NIH funding.

4.2 Evolution and Expansion of the Field of Women's Health Research

Due in great part to the concerns of scientists, health professionals, and advocates about a narrow focus on reproductive health, often to the exclusion of other important aspects of women's health, in the early 1990s the field was gradually redefined. The conceptualization of women's health began to include considerations of the lifespan continuum, especially focusing on the menopausal transition, issues of aging, post-menopausal health, factors contributing to the health and well-being of the frail elderly, and early life behaviors or contributors that could determine health status and risks for disease in later life. Many areas of reproductive health were determined still to be important for future research, including common conditions such as uterine fibroids and endometriosis, the effects of the intrauterine environment on adult health, gynecologic malignancies including breast, endometrial and ovarian cancers, and fertility and contraception. However, there was also a growing recognition of a need to determine sex and gender differences in the risk factors, diagnosis, response to treatment, and outcomes of health conditions beyond the reproductive system [21]. This resulted in an emphasis on the need for studies designed to determine when and to what extent these differences exist.

Recognizing that women from diverse populations may differ in response to interventions and health outcomes among themselves, just as they may differ from men, the women's health and research communities also began to focus on factors that contribute to disparities among different populations of women. These factors include psychological, behavioral, educational, social and non-social environmental influences, in addition to the effects of poverty, language barriers, and racial and cultural differences. Variations in access to health care and access to participation in clinical research studies were also given attention. NIH has increasingly initiated research on disparities among health outcomes of different populations and the translation of scientific findings to ensure improvements in health within communities and populations in need [22]. This public health focus on health disparities, as applied to women's health research, requires researchers to consider community, sociological and macro-environmental factors that influence health outcomes [23, 24]. There were also appeals for a less fragmented and more comprehensive approach to women's health care [21, 25–29].

4.2.1 Expanding the Vision for Women's Health Research

In 1997, ORWH undertook a second systematic program of strategic planning, which culminated in the publication of an *Agenda for Research on Women's Health for the 21st Century* [30]. The *Agenda* highlighted the potential for synergistic benefits of interdisciplinary women's health research and led to the development by ORWH of interdisciplinary funding initiatives. The first of these initiatives, intended to facilitate interdisciplinary women's health research career advancement, was modeled on an existing NIH program which had been designed to increase the number of physician scientists in women's reproductive health.

The *Women's Reproductive Health Research Career Development Centers* (WRHR) program was initiated in 1998, by the *Eunice Kennedy Shriver* National Institute of Child Health and Human Development (NICHD) in response to an identified shortage of obstetrician-gynecologist (OB/GYN) physician scientists conducting academic research in women's reproductive health [31]. The goal of the program was to encourage translational research in the areas of obstetrics and gynecology that would improve the health of women by advancing the research career development of OB/GYN investigators. The WRHR is an institutional, mentored research career development award designed to "build a national capacity of junior investigators in women's reproductive health research, provide junior faculty members with state-of-the art training in women's reproductive health research in an academic setting, stimulate women's reproductive health research over a variety of disciplines, and secure an outstanding research experience for junior faculty leading to a career as a successful independent investigator [32]". The program aimed to strengthen the research capacity in professional health institutions and meet the demand for scientists with a clinical background who are able to address increasing research opportunities in women's reproductive health. With NICHD as the lead institute and primary funding source for the program, ORWH participated in the development of the program and has been a contributor to the funding of the WRHR centers since its inception. As of 2010, there were twenty WRHR programs in OB/GYN Departments at academic institutions nationwide and a total of 157 OB/GYN researchers have completed or are currently enrolled in the program.

The WRHR provided an institutional model for an ORWH career development program that could attract and advance careers for doctoral level scientists in women's health research. A new program was conceptualized to diversify approaches to women's health research by providing an opportunity to include, not just reproductive health physician scientists, but also those from other medical professions as well as from many other health or related science disciplines [33]. Considering the much discussed need to overcome the fragmentation of women's health care, such a career development program also seemed an ideal way to promote a collaborative and comprehensive institutional approach to the investigation of common or complex problems by encouraging the creation of new relationships among diverse scientific disciplines.

4.2.2 Designing New Programs to Promote Interdisciplinary Women's Health Career Development and Research

With these principles in mind, in 1999 ORWH introduced the *Building Interdisciplinary Research Careers In Women's Health* (BIRCWH) Program to attract and prepare new investigators in women's health research, provide the mentoring and senior faculty support needed for their success, and deliver a new cadre of independently funded researchers experienced in multidisciplinary collegiality across investigative/professional settings [34]. In 2001 ORWH introduced a *Specialized Centers of Research on Sex and Gender Factors Affecting Women's Health* (SCOR) Program to support interdisciplinary research by accomplished scientists [35]. In 2007, the Office introduced a third interdisciplinary program, *Advancing Novel Science in Women's Health Research* (ANSWHR), to support investigator-initiated early stage interdisciplinary research on women's health and sex/gender factors [36, 37]. These programs have flourished and are now among the most successful NIH efforts to increase researchers in the field of women's health with an emphasis on determining sex and gender contributors to health using an interdisciplinary approach. The following section discusses current definitions of interdisciplinary research.

4.3 Interdisciplinary Research Defined

What is interdisciplinary research? Multiple definitions have been proposed [38]. When ORWH designed its interdisciplinary programs, the use of the term was based on the idea of combined efforts of various scientific disciplines working together with a blending of discrete areas of expertise to explore new scientific concepts.

In a 2004 National Academies (NA) Report *Facilitating Interdisciplinary Research*, interdisciplinary research was defined as a mode of research, often conducted by teams, that integrated information, data, techniques, tools, perspectives, concepts, and/or theories from two or more disciplines or bodies of specialized knowledge to advance fundamental understanding or to solve problems whose solutions are beyond the scope of a single discipline or area of research practice [39]. The NA definition distinguished between traditional multidisciplinary research and interdisciplinary research, with the key difference being that multidisciplinary research involved two or more disciplines collaborating by making *separate* contributions, whereas in interdisciplinary research, tools, methods, concepts and theories from two or more disciplines often result in an integration of discrete areas of expertise to explore new scientific concepts and may even lead to the emergence of new disciplines.

Interest in interdisciplinary approaches heightened with recognition that the most common diseases, such as heart disease, cancer, diabetes, or psychiatric illness, are also very complex. Their causation cannot be ascribed to a single gene mutation or one environmental factor; thus the study of these diseases often necessarily entails the consideration of multiple variables and advanced methodologies. A major challenge facing biomedical researchers is the need to sort out how multiple contributing factors interact and to translate this understanding into effective strategies for disease diagnosis, prevention, and therapy [40].

The NIH Roadmap for Medical Research was initiated in 2004 to address roadblocks to research and to transform the way biomedical research is conducted by overcoming specific hurdles or filling defined knowledge gaps [41]. The Roadmap is a collection of far-reaching initiatives designed to transform the Nation's medical research capabilities and improve the translation of research into practice. The Roadmap consists of three major themes: (1) new pathways to discovery; (2) research teams of the future; (3) and reengineering the clinical research enterprise. The initiatives within the theme of medical discovery aim to improve the understanding of complex biological systems and to develop tools that will advance biomedical research. For the future research team, the initiatives are designed to encourage scientists and scientific institutions to test a variety of models for conducting research. Furthermore, the initiatives for the clinical research enterprise are central to the goal of moving research results more quickly into clinical settings [42–44].

Under the elements of the Roadmap definition, interdisciplinary research requires the bridging of traditional intellectual and structural divisions between scientific disciplines to speed the pace of scientific discovery. Interdisciplinary research integrates analytical strengths of two or more disparate scientific disciplines to eventually create a new hybrid discipline. This vision required new organizational models of research. Traditional biomedical research organizations, structured much like a series of cottage industries, separated out researchers into broad areas of scientific interest and then further subdivided them into distinct, departmentally based specialties.

The ORWH implementation of interdisciplinary research programs preceded the NA and NIH Roadmap definitions and had a specific focus on the needs of the field of women's health. Interdisciplinary women's health research was viewed as a means to facilitate the integration of diverse scientific approaches and to meld the expertise and differing perspectives of health professional disciplines and specialties, basic science fields of study, and other potential contributing areas such as engineering, mathematics, public health, anthropology, social science, biotechnology, or additional fields that could contribute to fresh insights and mutually beneficial collaborations.

4.4 Building Interdisciplinary Research Careers in Women's Health (BIRCWH) Program

Utilizing the concepts of the WRHR centers, ORWH introduced a new program dedicated to training and advancing the careers of researchers in the conduct of interdisciplinary women's health research in a strongly mentored environment. The BIRCWH program supports the mentored research career development of junior faculty members, known as BIRCWH Scholars, who have recently completed clinical training or postdoctoral fellowships, and who would engage in interdisciplinary basic, translational, behavioral, clinical, and/or health services research relevant to women's health or sex/gender factors. The program was built around three pillars: interdisciplinary research, mentoring and career development. BIRCWH is a trans-NIH collaborative effort, which is supported by many NIH ICs, the NIH Office of Dietary Supplements, and the Agency for Healthcare Research and Quality (AHRQ). The program was designed to increase the number and skills of investigators at institutions through a 2–5-year mentored research experience during which senior investigators serve as mentors to BIRCWH Scholars. The mentors at each site are investigators committed to fostering interdisciplinary approaches to research in women's health and sex/gender factors and to providing advice to the Scholars on their career advancement and research. Because of the interdisciplinary nature of this career development program, the mentors must represent diverse areas of expertise. The first BIRCWH sites were funded in 2000 and as of 2010, fifty awards have been granted to thirty-eight research universities across the country.

A total of 378 individuals have participated as BIRCWH Scholars, generally with four Scholars per program, per year, as of November 2009. Although the program invites the participation of both men and women, 79% of BIRCWH Scholars have been female. The program is open to Scholars in diverse disciplines or specialties, who have a doctoral degree and no more than 6 years of subsequent research training. Among all BIRCWH Scholars, 30% have a PhD; 25.1% have an MD; 12.7% have both an MD and a PhD; and 8.5% have both an MD and an MPH. Other Scholars hold nursing, engineering, physical therapy or business degrees, doctorates in pharmacy, or other degrees. Among Scholars, the two most frequently reported disciplines are internal medicine (22.7%) followed by obstetrics/gynecology (10.7%). BIRCWH Scholar research spans the full spectrum of women's health topics and includes attention to a range of health conditions, health services, trauma, and health disparities. Table 4.1 provides a listing of the most common research areas identified by BIRCWH Scholars.

Because of the importance of mentoring for successful career advancement in research and for navigating the academic environment for promotions and faculty expectations, mentoring is an essential component of the BIRCWH program. While participating in the BIRCWH program, Scholars receive mentoring from senior faculty members from at least two disciplines and/or fields, and they are provided with 75% protected time to do their research. An annual meeting at NIH of all BIRCWH Scholars provides an opportunity for them to share their mentoring experiences and to learn about models of mentoring in other programs that may enhance those at their home institutions. The list of benefits of mentoring, as seen in Table 4.2, identified by a Scholar from the University of North Carolina BIRCWH program and presented at the 2007 NIH National Leadership Workshop on Mentoring Women in Biomedical Careers, provides a virtual list of ideas for which mentors can be of great value in guiding the careers of investigators [45].

A major purpose of the BIRCWH program is to provide opportunities for a mentored experience that will help to bridge the transition to research independence for junior faculty researchers. One measure of its success in achieving this goal is the number of scholars who apply for and receive NIH or other funding as independent investigators. As of 2010, 69% of BIRCWH Scholars who successfully completed the program reported submitting at least one NIH grant application while still in the program and 44% were successful in receiving at least one independent NIH research grant, for a total of 282 grants. Scholars have also received Federal funding from AHRQ, the Centers for Disease Control and Prevention, the Department of Defense and the Department of Energy. Scholar research has also been supported through a variety of academic, foundation and industry grants. Another measure of success is the number of scholarly publications produced by Scholars. As of 2010, BIRCWH Scholars have reported over 1,300 resulting publications.

Table 4.1 Examples of BIRCWH scholar research areas

• Mental health	• Menopausal hormone therapy
• Diabetes	• Sex differences research
• Cardiovascular health	• Substance abuse
• Neurological disorders	• Cancer research
• Trauma	• Molecular biology/genetics
• Reproductive health	• Health services research
• Obesity	• Arthritis/musculoskeletal health
	• Health disparities research

Table 4.2 The benefits of mentoring as described by University of North Carolina BIRCWH scholar Julia E. Britain, Ph.D. at the NIH national leadership workshop on mentoring women in biomedical careers

How to write	How to balance	How to tolerate IRBs
How to speak	How to create a positive place	How to cut through red tape
How to be successful	How to build a team	How to really take action
How to handle failure	How to be fearless	How to challenge paradigms
How to persevere	How to publish	How to get out of a bad spot
How to do research	How to be absolutely still	How to transition your career
How to teach someone to do research	How to respond in a crisis	How to go where the data take you
How to be a mentee	How to recognize a crisis	How to use the administration
How to go after what I want	How to avoid a crisis	How to not go it alone
How to conduct myself	How to pick a good mentor	How to know this is up to me
How to stay positive	How to say NO	How to end a project
How to shut up	How to say not now	How to be patient
How to ask for advice	How to say yes	How to push and push and push
How to follow advice	How to let go of the past	How to recruit people
How to use time	How to set goals for the future	How to prioritize
How to love what I do	How to take responsibility	How to use punctuation
How to think	How and when to take control	How to proofread
How the data usually don't lie	How to get a free lunch	How to lead
How to make others successful	How to ask the best questions	How to read and read and read
How to take the high road	How to interact with colleagues	How to be political
How to handle politics	How to make people think about me	How to take credit
How to promote myself	How to get left alone	How to stand up for those that can't
How to invest in an individual	How to hire someone really good	How to simply be the best.
	How to take the time I need	
	How to tolerate incompetence	

4.5 Specialized Centers of Research on Sex and Gender Factors Affecting Women's Health (SCOR)

The SCOR program was the second major interdisciplinary research initiative developed and implemented by the ORWH. Following the enthusiasm with which the BIRCWH had been received, the SCOR was developed to support the initiation and expansion of interdisciplinary collaboration in research among established researchers.

At the time that the SCOR program was conceptualized, much new attention was being given to the importance of addressing sex differences and the role of biological sex in basic research. A policy for the inclusion of women in clinical research in NIH-funded studies existed, but no policy was in place to encourage considerations of factors related to biological sex in basic research. In 2001, the Institute of Medicine released the report, *Exploring the Biological Contributions to Human Health: Does Sex Matter?*, which emphasized that multiple sex and gender factors needed to be considered in research [46]. Furthermore, a failure to consistently distinguish between gender and sex factors influencing women's health was a conceptual problem that could impede progress in the field. The report noted that "an additional and more general reason for studying differences between the sexes is that these differences, like other forms of biological variation, can offer important insights into underlying biological mechanisms." In order to better understand human health, this report also encouraged that research should determine natural variations with regard to sex/gender in genetic factors, disorders of sexual differentiation, reproductive status, and environmental influences.

With the new attention to the role of sex/gender factors in human health from the molecular level to clinical interventions, it seemed timely to implement an interdisciplinary program focused specifically on sex and gender factors. The SCOR program, which was first funded in 2002, was designed to stimulate the spectrum of interdisciplinary sex/gender research from the basic level to its translation into clinical applications. Each SCOR is required to have a central theme related to the disease area to which the individual projects relate and which serves as an integrating force. A SCOR must also provide an interdisciplinary approach utilizing both

Table 4.3 ORWH specialized centers of research on sex and gender factors affecting women's health, 2007–2011

Jill Goldstein, Ph.D., Brigham and Women's Hospital
Fetal antecedents to sex differences in depression: a translational approach

Kathleen Brady, M.D., Ph.D., Medical University of South Carolina
Role of sex and gender differences in substance abuse relapse

Andrea Dunaif, M.D., Northwestern University
Excess male hormones (androgens) as the key to explaining polycystic ovarian syndrome

Emeran Mayer, M.D., University of California, Los Angeles
A coordinated study of stress, pain, emotion, and sexual factors underlying the pelvic visceral disorders of irritable bowel disorder and interstitial cystitis

Jeanette Brown, M.D., University of California, San Francisco
Lower urinary tract function in women

David Ehrmann, M.D., University of Chicago
Sex steroids, sleep, and metabolic dysfunction in women

Emmalee Bandstra, M.D., University of Miami
Sex and gender influences on addiction and health: a developmental perspective

John DeLancey, M.D., University of Michigan, Ann Arbor
Birth, muscle injury and pelvic floor dysfunction

Hong-Wen Deng, Ph.D., University of Missouri, Kansas City
Identifying the genes that put women at risk for osteoporosis

Scott Hultgren, Ph.D., Washington University
Molecular and epidemiologic basis of acute and recurrent urinary tract infections (UTIs) in women

Rajita Sinha, Ph.D., Yale University
Sex, stress, and substance use disorders

laboratory and clinical research to focus on a particular health problem and provide for a mutually supportive interaction between basic scientists and clinical investigators. Integral components of projects include the interdisciplinary development of innovative approaches, elaboration of new and significant hypotheses, and generation of improved strategies for approaching current issues relating to the disease area addressed. A SCOR may be interdepartmental, intercollegiate, or inter-institutional.

A SCOR consists of at least three individual, but interrelated, research projects, each with high scientific merit and clear research objectives and, in the aggregate, devoted to a specific major health area. Both basic and clinical research must be present. An administrative core coordinates the research program, providing intellectual leadership as well as basic management functions, and one or more resource cores may also be present. Cores provide resources to be shared by the multiple investigators working on the individual SCOR research projects. The cores enhance research productivity and increase the functional capacity of the SCOR.

Since the program first provided funding in 2002, a total of twenty-two SCOR awards have been made to fifteen institutions across the US. Table 4.3 lists currently funded SCORs and their primary areas of focus.

4.6 Advancing Novel Science in Women's Health Research (ANSWHR) Program

In an effort to increase investigator interest in women's health research and sex/gender research, ORWH and multiple co-sponsoring NIH ICs initiated the ANSWHR program in 2007. The program, which funds investigator-initiated research grants, aims to promote early stage developmental research that will advance innovative interdisciplinary concepts in women's health research and the study of sex/gender factors. Examples of areas of science that have been funded through the program include: cancer, human genetics, heart disease, immunology, lupus, infectious diseases, mental health, pain, neurology, obesity, pulmonary dysfunction, pregnancy, ovarian dysfunction, HIV/AIDS, drug and alcohol abuse, and stress.

4.7 Interdisciplinary Research and Career Development: Benefits and Challenges

The NIH interdisciplinary programs in women's health research were initiated as a new paradigm to move forward the science associated with women's health and to better understand the contributions of sex/gender to human health and disease. Starting with the BIRCWH program to attract future investigators to consider women's health research and to sustain their interest as research scientists through a mentored career development program, this endeavor has continued and expanded because of the reported progress of the Scholars and their enthusiasm. The interdisciplinary aspects have afforded new models of mentoring by senior scientists from two or more disciplines, providing enhanced career and scientific guidance to the Scholars to further their academic pursuits as well as their research skills. Comments from Scholars about the value of having protected time as well as guidance in their research have been complimentary, and individual programs have developed a variety of approaches to interdisciplinary mentoring from team and group meetings to scheduled individual one-on-one sessions. Table 4.4 contains examples of such comments.

The success rate for Scholars' applications for NIH funding has been encouraging; a complete evaluation of first time applications and continued success rates is pending. Further follow up of academic success and subsequent interdisciplinary pursuits is also under study. However, the number of publications by BIRCWH Scholars emphasizes that, although this is a career development program, these Scholars are making progress in their scientific endeavors. The annual interdisciplinary program meetings at the NIH include all BIRCWH Scholars and program directors, as well as the principal investigators of the SCORs. During this meeting, an annual Interdisciplinary Women's Health Research Symposium is held to feature progress by BIRCWH Scholars and SCOR investigators [47–50]. Four sets of meeting abstracts have been published [51–54]. The science has been exciting and the resulting networking among those with similar investigative interests is leading to inter-institutional collaborations. The SCOR, BIRCWH, and ANSWHR programs complement other Federally-supported programs addressing women's health issues and career development awards, such as the WRHR and other mechanisms.

The potential positive impact of furthering institutional scientific endeavor across disciplines was reported by Hultgren and Elam [55]. After they received one of the first interdisciplinary SCOR grants to study the molecular and epidemiological basis of acute and recurrent urinary tract infections in women, their initial progress resulted in the creation of a new Center for Women's Infectious Disease Research

Table 4.4 Comments from Washington University BIRCWH scholars regarding the impact of BIRCWH submitted at the *moving into the future – new dimensions and strategies for women's health research for the National Institutes of Health* public hearing on March 4, 2009 in St. Louis, MO

"The financial support provided to me as a BIRCWH scholar quite literally saved my research career... I was working on a research project that represented a completely new avenue of research for my Principal Investigator and me; my progress had been slow... Receipt of BIRCWH funding enabled me to continue to pursue this line of investigation, and provided me with the time to obtain necessary data and publications." – Lyse Norian, Ph.D.
"The BIRCWH program allowed for me to have a significant amount of protected time and the resources to successfully begin my research career... One of the other major benefits of the BIRCWH program was ... visiting professors in my field ...had to spend time with me discussing my chosen field of research – obesity-related heart disease. This networking and interacting with other researchers in my field was invaluable to my career." – Linda Peterson, M.D.
"I now appreciate why the BIRCWH program places such an emphasis on mentorship. Simply stated, it is essential." – Joan Riley, Ph.D.

at Washington University in St. Louis which will study the impact of sex and gender issues on susceptibility to diseases that are caused by microbial pathogens. This center, which received institutional support and the collaboration of three departments (Microbiology, Medicine and Biochemistry), as part of this new university initiative, will study the microbial pathogenesis of such conditions as "sexually transmitted diseases, infections that lead to complications in pregnancy, interstitial cystitis and painful bladder syndrome, vulvodynia, cancers that are associated with infective agents, such as cervical cancer, and the possible roles of infection in other diseases, such as neurodegenerative disorders, heart disease and autoimmune disorders." The investigators acknowledge that new interdisciplinary centers such as that established at Washington University will require the development of a different type of institutional infrastructure.

Institutional challenges to implementing interdisciplinary research programs have been reported by others. Investigators of the University of Michigan BIRCWH program, with participation of four schools (Medicine, Nursing, Public Health and the College of Literature, Science and the Arts) describe practical barriers such as difficulties organizing meetings, developing a common language and knowledge and understanding of the tasks to be undertaken [56]. Perhaps of more consequence is that most academic institutions are organized by disciplines or dedicated departments. Therefore, investigators who are engaging in interdisciplinary research efforts may have to creatively overcome the traditional organizational infrastructure and disciplinary approaches to allocation of resources, promotion policies, attribution on publications, and dedicated didactic perspectives. Instituting trans-departmental and even trans-university models of successful collaboration is important if the intended nature of multidisciplinary cohesion is to be achieved in addressing complex problems through a unity of effort.

When the first women's health interdisciplinary programs began, concerns were expressed about peer review for future investigator initiated interdisciplinary applications. However, as other interdisciplinary research initiatives expanded, the NIH Center for Scientific Review added a new Division of Translational and Clinical Sciences to accommodate the increasing number of interdisciplinary grant applications [57]. As interdisciplinary research and career development programs advance, those who are engaged in them may also develop another dimension of importance to academic clinical departments in providing clinicians-in-training with knowledge of the multiple scientific perspectives about the pathogenesis of and approach to conditions falling under their disciplinary clinical purview, and providing them with appreciation of the interrelatedness to other conditions affecting their patients' health, enhancing the comprehensive approach to women's health and health care.

4.8 Conclusions and New Models of Women's Health Care

In this era of expanded concepts of women's health and health care, it is anticipated that demand for providers who can treat the "whole woman" will increase [58]. Traditionally, patients and health professionals have functioned within a delivery model that encourages patients to seek specialized care from a variety of professionals. Designing and implementing a new model of women's health care will require a conceptualization of care focused on comprehensive approaches. A number of model programs have been developed [59–61]. These authors concluded that interdisciplinary models can provide effective training for post-graduate physicians and enable the provision of integrated care to women, but also that their success depends on highly developed collaborative relationships between faculty, nonclinical sources of support, and long-term institutional commitment.

Interdisciplinary women's health research arose out of a perceived need for the sort of integrated lifespan research that includes multiple perspectives. Interdisciplinary research teams can serve as models of the kinds of collaborations that will be required for successful non-fragmented women's health care. The demand for physicians, regardless of their specialty, to serve as *de facto* primary care doctors for women is a reality [62]. The recommended content of women's primary and preventive care is articulated in several sources, including the US. Preventive Services Task Force Guide to Clinical Preventive Services [63] and the American College of Obstetricians and Gynecologists' Guidelines for Women's Health Care, Third Edition [64]. This emphasis requires new perspectives which encompass many aspects that contribute to health and wellness from the molecular and biologic bases of diseases to lifespan considerations, social and societal factors, health disparity perspectives, and public health implications. Finally, an increased demand for personalized medicine – medicine that takes into account sex and gender differences in risk, incidence, diagnosis, course, treatment response and outcome – can benefit from the findings from translational and interdisciplinary research teams. Speeding translation requires a steady pipeline of basic, clinical and translational researchers; interdisciplinary teams that can collaborate with federal, industry, and community partners; and integrated resources and informatics to achieve cost efficiencies that will yield new products, approaches, and diagnostic tools in less time [43].

4.8.1 New Dimensions and Strategies for Women's Health Research at the National Institutes of Health

Just as interdisciplinary research has blossomed during the last decade into a major component of women's health research, new paradigms and new programs to advance women's health research continue to be developed. Ten years after the last women's health research agenda was updated, the NIH has launched a series of regional scientific workshops and public hearings to ensure that research on women's health continues to be on the cutting edge of science and is based upon the most advanced and anticipated techniques and methodologies. Individuals representing the full spectrum of academic institutions, professional associations, advocacy organizations and healthcare facilities interested in biomedical and behavioral research on women's health and sex/gender issues are contributing to the consideration of new directions for future studies. Considerations include approaching women's health from the totality of the female life span including the prenatal, infant and childhood, adolescent, reproductive, menopausal and elderly years. The role of sex and gender factors in all aspects of a girl or woman's self-identification, health, and health care are central to future designs of investigative efforts. As the 21st century progresses, research will focus on how emerging fields such as systems biology, stem cell and genetic research, new technologies such as imaging and engineering advances, and environmental factors may be applicable to women's reproductive and overall health.

Priorities for the next decade of research will emphasize continuing gaps in knowledge as well as emerging questions based upon recent research revelations, with implications for basic research as well as clinical translation and public health policies. Lifespan considerations will encourage such issues as how the intrauterine environment and conditions experienced during pregnancy affect health in later life and aging. Research has typically focused on abnormal physiology and disease, leaving gaps in the understanding of normal physiology and health maintenance. There are recommendations for a systems biology approach to women's diseases and co-morbidities, in particular for addressing older women who are more likely to be taking multiple medications to treat multiple conditions. Such studies are being suggested to provide a better understanding of drug interactions and the comparative effectiveness of interventions.

Development of new technologies and novel applications of existing technologies, including advances in biomedical imaging and microbiology techniques, high-throughput synthesis and screening of small drug-like molecules, genetic sequencing and nanotechnology are recommended to be incorporated into interdisciplinary studies of women's health and sex/gender factors. Development of minimally invasive technologies that can capitalize on the intersections of various technologies can be used to diagnose diseases at early stages, deliver treatments and improve outcomes. The expanded application of other technologies, such as telemedicine and point of care diagnostics, are suggested to assist in translating new knowledge into all clinical settings, including those that continue to be undeserved such as rural or urban areas and underdeveloped countries.

As women's health research and health care increase in complexity and the expectations of patients become more informed, it is increasingly important to encourage scientists with diverse expertise to consider sex and gender in the design of their studies. The resulting advances in knowledge can then enable the application of sex specific personalized medicine of importance to the health of women and men. During the past decade, the WRHR, BIRCWH, and SCOR programs have contributed to the attraction and retention of research scientists in women's health and the ANSWHR program has provided an additional mechanism for funding the development of innovative concepts. Continuing to expand the cadre of physician scientists with an understanding of interdisciplinary and comprehensive approaches to health and disease will be crucial for new strategies in women's health research and their translation into practice in future years; novel and successful initiatives to promote these careers must continue.

4.8.2 Women in Biomedical Careers – The Changing Landscape in Medical and Research Careers in the United States

The year 2010 marks 100 years since Abraham Flexner issued his report to the Carnegie Foundation on *Medical Education in the United States and Canada* and reported that in 1909, there were 921 women medical students in 94 schools with 162 graduates. He observed that, "Now that women are freely admitted to the medical profession, it is clear that they show a decreasing inclination to enter it. More schools in all sections are open to them; fewer attend and fewer graduate. . ." [65]

Flexner's conclusions that women had a declining interest in becoming physicians at that time did not identify other factors that might have served as barriers to women's entry into the profession. Over the past half century, that there has been a striking increase in the numbers of women applicants to medical schools in the United States, and of those admitted and graduating. By academic year 2008–2009, approximately 48% of applicants accepted to US. medical schools were women, and 49% of graduates were women, as is demonstrated in Fig. 4.1 [66].

Consequent to the increasing numbers of women medical school graduates there has been an increasing proportion of women among residents. Figures 4.2 and 4.3 illustrate that in 2008, women constituted the highest proportion of residents in obstetrics and gynecology (78%), pediatrics (69%) and medical genetics (66%), in contrast to men who were the highest proportion of residents in neurological surgery (88%), orthopedic surgery (87%), thoracic surgery (87%), and urology (78%). Recently, however, the largest increases in the proportion of women have

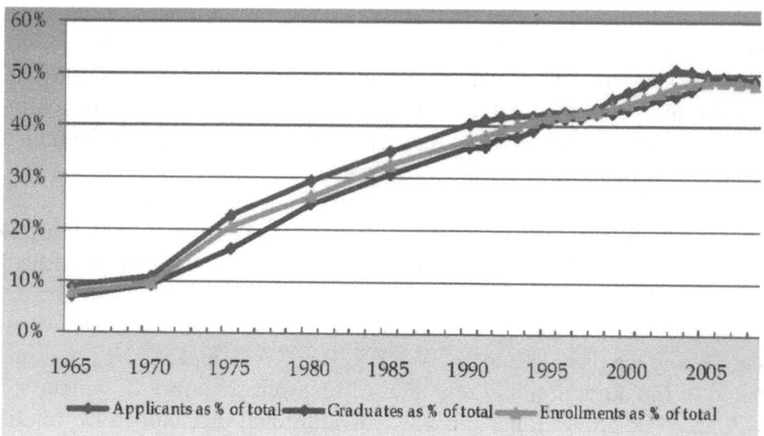

Fig. 4.1 Changes in women enrollment, applicants & graduates, 1965–2009.
As displayed in Figure, the most notable changes in women representation among medical students occurred between 1970–1971 and 1990–1991. In those 20 years, the proportion of women applicants and graduates increased from 11 to 40%, and 9 to 36%, respectively.

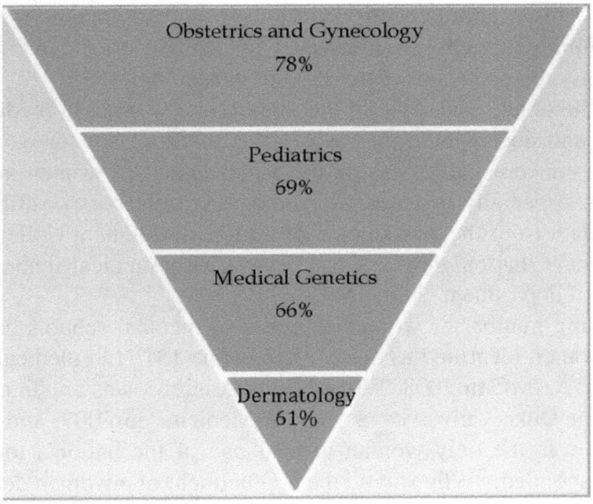

Fig. 4.2 Specialties with the highest proportion of women residents, 2008.
As shown in Figure, obstetrics and gynecology (78%), pediatrics (69%) and medical genetics (66%) lead as the specialties with the highest proportion of women residents, as compared to men. Among all women residents, the top four selected specialties were: internal medicine (20%), pediatrics (16%), family medicine (11%), and obstetrics and gynecology (8%).

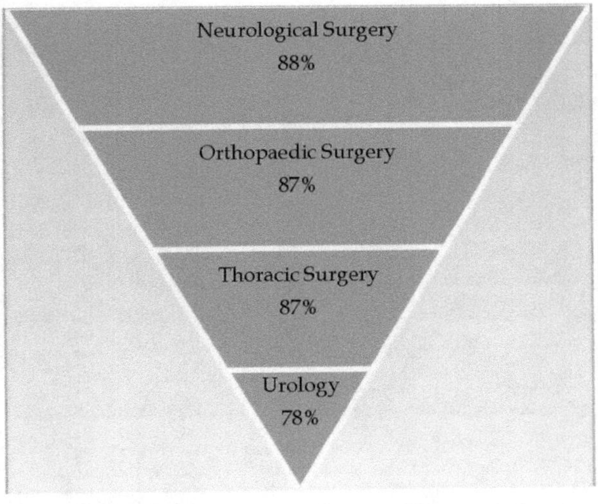

Fig. 4.3 Specialties with the highest proportion of men residents, 2008.
As shown in Figure, the specialties with the highest proportion of men residents in 2008 were: neurological surgery (88%), orthopedic surgery (87%), thoracic surgery (87%) and urology (78%). Of all men residents, the top four selected specialties were: internal medicine (21%), internal medicine subspecialties (11%), surgery (9%), and family medicine (7%).

been in residencies in thoracic surgery, urology, and orthopedic surgery, but the residency training programs most selected by women are in internal medicine, pediatrics, family medicine and obstetrics and gynecology [66].

Even with the notable increase in the numbers of women in medicine over the past 40 years, and documented progress in the opportunities for women in medical disciplines, concerns still focus on the advancement of women in medical and science careers, especially in academic medicine or university positions in science [67]. Current data from the Association of American Medical Colleges, shown in Fig. 4.4, document that only 18% of full professors in medical schools are women and only 13% of department chairs [66].

The increasing number of women as deans of medical schools has progressed slowly, with women identified as deans at 16 of the 131 U.S. medical schools represented by the AAMC in 2008. When Nancy Andrews was appointed as the first woman Dean of Duke University School of Medicine in 2007, and news reports noted that she was the only woman to lead one of the nation's top 10 medical schools, she responded, "Why should the appointment of a woman dean still be big news in 2007? Perhaps because, with a few localized exceptions, there has been little change since the 1970's in the barriers to women's full participation in academic medicine [68]".

Much attention has been given to the barriers to the successful advancement of women in biomedical and academic careers, resulting in what some refer to as the "leaky pipeline" [9, 69–72]. A 2007 National Academies report, *Beyond Bias and Barriers: Fulfilling the Potential of Women in Academic Science and Engineering*

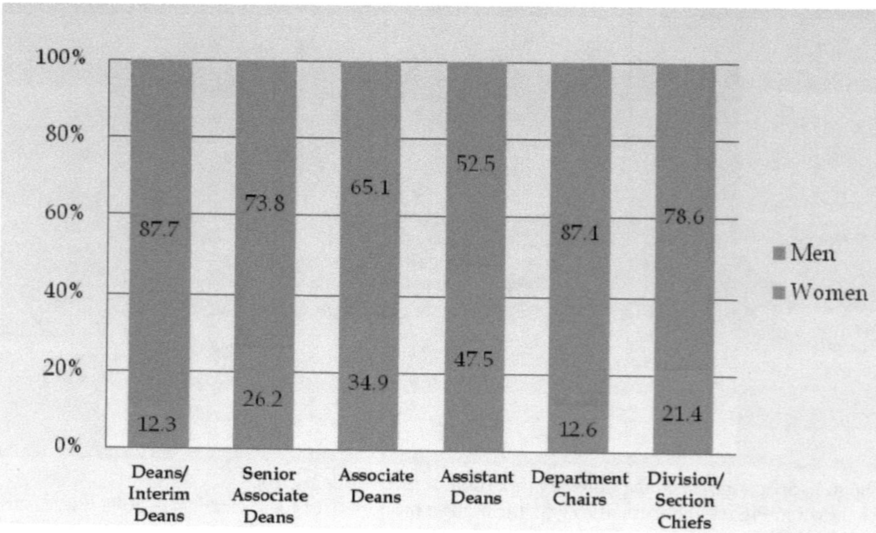

Fig. 4.4 Leadership at US. medical schools by gender, 2008.

[73] provides an in depth review of the issues confronting women in academic biomedical careers. This report concluded that, contrary to some suggestions that women may lack the innate ability to achieve in science careers, "It is not lack of talent, but unintentional biases and outmoded institutional structures that are hindering the access and advancement of women." This report further emphasized the importance of an environment that will permit women to realize their full intellectual potential in their careers, and it challenged research funding and federal agencies and foundations to ensure that their practices and rules do not reinforce institutional biases that could discriminate against women.

The NIH responded to this challenge through the establishment of a trans-NIH Working Group on Women in Biomedical Careers, led by the NIH Director and the Director of the Office of Research on Women's Health [74]. This Working Group is developing innovative strategies and tangible actions by which the NIH can promote the advancement of women in research or academic careers [75, 76]. It is examining the effectiveness of programs to reduce bias, including a funding initiative to study the efficacy of interventions that promote research careers to better design future innovations [77, 78].

The Working Group has clarified federal cost policies to provide guidance to universities on how to structure their indirect cost and fringe benefit arrangements when considering reimbursement for child care, parental leave and other factors in response to some of the most frequently asked questions for investigators with family responsibilities [79]. Considering the need to assist dual-career couples to find employment near each other, NIH helped establish the Mid-Atlantic Higher Education Recruitment Consortium, which provides networking between regional universities and agencies, and web-based resources about job opportunities [80].

A comparison of success rates of women and men applying for NIH grant support yields interesting data [81]. Experienced males have higher success rates than females but the gap is narrowing although not closing, but first-time male and female researchers are equally successful. Both sexes receive about the same percentage of their requests but the average female request for research project grants is less than that of the average male request, therefore on average, males receive larger awards than females.

Because mentoring is crucial for sustaining biomedical careers, the NIH National Leadership Workshop on Mentoring Women in Biomedical Careers was held to identify innovative and effective mentoring programs and practices. Examples of different paradigms for mentoring included institutional "cross-mentoring" between disciplines, and "panel mentoring," where one mentee can call on more than one mentor for advice. Incorporation of mentoring into tenure-and-promotion reward systems would lead to a system where mentoring is not only expected, but is valued [45].

Deliberations on best practices for sustaining career success for women in biomedical careers were held during a NIH national workshop designed to provide examples of systems approaches used by businesses, military and academic health centers, as well as NIH, to ensure sustained career success. Among the ideas proposed for organizations to address in policy or actions were securing a strong personal commitment to change from top management; employing open and transparent policies for hiring, salary and promotion decisions; instituting "family-friendly" policies; and, providing fair and equitable access to physical, financial and organizational resources [82].

With concerns about increasing opportunities in and advancing biomedical careers for women and men, ORWH supported interdisciplinary career development programs was developed, and has added another dimension to potentiating research and academic careers in medicine. Looking back on the comments of Flexner 100 years ago, the advent of organizational activities that promote career development for women and men, but especially those that are designed to overcome institutional policies and programs that deliberately or incidentally promote gender bias, can expedite and sustain progress and parity in medical and research careers for the coming years.

Acknowledgements For their valuable assistance in the preparation of this manuscript: Janine A. Clayton, M.D., Deputy Director, ORWH; Nida H. Corry, Ph.D., AAAS Science & Technology Policy Fellow, ORWH; Joslyn Yudenfreund Kravitz, Ph.D., Special Assistant to the Director, ORWH; and Joan Davis Nagel, M.D., M.P.H., Director of Interdisciplinary Research Programs, ORWH.

References

1. Legato, M.J. 2000. Is there a role for gender-specific medicine in today's health care system? *Journal of Gender Specific Medicine* 3(12):15–21.
2. Legato, M.J. 2003. Beyond women's health: The new discipline of gender-specific medicine. *Medical Clinics of North America* 87(5):917–937.
3. Legato, M.J. 2006. Summary. *Journal of Women's Health* 2(5):699–703.
4. Wenger, N.K. 2004. You've come a long way, baby: Cardiovascular health and disease in women: Problems and prospects. *Circulation* 109:558–560.
5. Raymond, S.U., H.M. Greenberg, and S.R. Leeder. 2005. Beyond reproduction: Women's health in today's developing world. *International Journal of Epidemiology* 34(5):1144–1148.
6. Donohue, G., ed. 1996. *Women's health in the curriculum: A resource guide for faculty*. Philadelphia, PA: National Academy on Women's Health Education.
7. Hamilton, J., M. Jensvold, E. Rothblum, and E. Cole., eds. 1995. *Psychopharmacology from a feminist perspective*. Binghamton, NY: The Haworth Press.
8. Mastroianni, A.C., R. Faden, and S. Federman., eds. 1994. *Committee on the ethical and legal issues relating to the inclusion of women in clinical studies, institute of medicine*. Washington, DC: National Academy Press.
9. Carnes, M., C. Morrissey, and S.E. Geller. 2000. Women's health and women's leadership in academic medicine: Hitting the same glass ceiling? *Journal of Women's Health* 17(9): 1453–1462.
10. 1976. The multiple risk factor intervention trial (MRFIT). *JAMA* 235(8):825–827.
11. Neaton, J.D., R.H. Jr Grimm, and J.A. Cutler. 1987. Recruitment of participants for the multiple task factor intervention trial (MRFIT). *Control Clinical Trials* 8(4 Suppl):41S–53S.
12. 1988. The physicians' health study: Aspirin for the primary prevention of myocardial infarction. *New England Journal of Medicine* 318(14):924–926.
13. Ridker, P.M., N.R. Cook, I.M. Lee et al. 2005. A randomized trial of low-dose aspirin in the primary prevention of cardiovascular disease in women. *New England Journal of Medicine* 352:293–304.

14. U.S. Public Health Service. 1985. Women's health: Report of the public health service task force on women's health issues. *Public Health Reports* 100(1):73–106.
15. 1987. NIH guide to grants and contracts. 16(3):2, 23 Jan 1987.
16. Nadel, M.V. 1990. *National institutes of health: Problems in implementing policy on women in study populations*. Washington, DC: General Accounting Office.
17. NIH/HHS News Release. September 10, 1990.
18. Office of Research on Women's Health, National Institutes of Health. 1992. *Report of the National Institutes of Health: Opportunities for research on women's health, volumes. 1 and* 2. Bethesda, MD: National Institutes of Health.
19. Public Law 103–43. National Institutes of Health Revitalization Act of 1993. 42 U.S.C. 289(a)(1).
20. 1994. NIH guidelines on the inclusion of women and minorities as subjects in clinical research. *Federal Register* 59:14508–14513.
21. Wallis, L., ed. 1997. *Textbook of women's health*. New York, NY: Lippincott-Raven.
22. Ruffin, J. 2010. A message from the director. National Center on Minorities and Health Disparities, National Institutes of Health. http://www.ncmhd.nih.gov/about_ncmhd/welcome.asp. Accessed 25 Jan 2010.
23. Satcher, D. 2001. American women and health disparities. *Journal of American Medicine Women's Association* 56(4):131–132.
24. Kaiser Family Foundation. 2009. *Putting women's health care disparities on the map: Examining racial and ethnic disparities at the state level,* June 2009.
25. Clancy, C.M., and C.T. Massion. 1992. American women's health care: A patchwork quilt with gaps. *JAMA* 268:1918–1920.
26. Hoffman, E., and J. Johnson. 1995. Women's health and enhancing generalism in internal medicine. *Journal of General Internal Medicine* 10:451–454.
27. Pinn, V.W. 2004. The view from the National institutes of health: A decade with the office of research on women's health. *Gender Medicine* 1(1):5–7.
28. Pinn, V.W. 2005. Interdisciplinary research on women's health and sex/gender factors. *Gender Medicine* 2(3):121–123.
29. Pinn, V.W. 2005. Research on women's health: Progress and opportunities. *JAMA* 294(11):1407–1410.
30. Office of Research on Women's Health, National Institutes of Health. 1999. *Agenda for research on women's health for the 21st century: A report of the task force on the NIH women's health research agenda for the 21st century*. Bethesda, MD: National Institutes of Health.
31. Institute of Medicine. 1992. *Strengthening research in academic OB/GYN departments*. Washington, DC: National Academy Press.
32. Women's Reproductive Health Research Career Development Centers RFA HD-98-004. NIH guide, 9 Jan 1998. http://grants.nih.gov/guide/rfa-files/RFA-HD-98-004.html. Accessed 25 Jan 2010.
33. Pinn, V.W. 2007. Introduction: Second NIH interdisciplinary women's health research symposium. *Journal of Women's Health* 15(6):669–670, 20 Oct 2005.
34. Building Interdisciplinary Research Careers in Women's Health, RFA-OD-99-088. NIH guide, Sept 7 1999. http://grants2.nih.gov/grants/guide/rfa-files/RFA-OD-99-008.html.
35. Specialized Centers of Interdisciplinary Research (SCOR) on sex and gender factors affecting women's health (P50) RFA-OD-02-002. NIH guide 18 Dec 2001. http://grants.nih.gov/grants/guide/rfa-files/RFA-OD-02-002.html.
36. Advancing Novel Science in Women's Health Research (ANSWHR) (R21) PAS-07-381. NIH guide, 11 Jun 2007. http://grants.nih.gov/grants/guide/pa-files/PAS-07-381.html.
37. Advancing Novel Science in Women's Health Research (ANSWHR) (R03) PAS-07-382. NIH guide, 11 Jun 2007. http://grants.nih.gov/grants/guide/pa-files/PAS-07-382.html.
38. Pellmar, T.C., and L. Eisenberg, eds. 2000. *Bridging disciplines in the brain, behavioral, and clinical sciences*. Washington, DC: National Academy Press.

39. National Academies. 2004. *Facilitating interdisciplinary research*. Washington, DC: National Academy Press.
40. Kibertstis, P., and L. Roberts. 2002. Introduction to special issue: It's not just the genes. *Science* 296:685.
41. NIH Roadmap. http://nihroadmap.nih.gov. Accessed 20 Jan 2010.
42. Zerhouni, E.A. 2005. Translational and clinical science – time for a new vision. *New England Journal of Medicine* 353(15):1621–1623.
43. Zerhouni, E.A. 2007. Translational research: Moving discovery into practice. *Clinical Pharmacology Therapeutics* 81(1):126–128.
44. Woolf, S.H. 2008. The meaning of translational research and why it matters. *JAMA* 299(2):211–213.
45. Office of Research on Women's Health, National Institutes of Health. 2008. *National leadership workshop on mentoring women in biomedical careers*. Bethesda, MD.
46. Institute of Medicine. 2001. *Exploring the biological contributions to women's health: Does sex matter?* Washington, DC: National Academy Press.
47. Pinn, V.W. 2007. Introduction: Second NIH interdisciplinary women's health research symposium October 20, 2005. *Journal of Women's Health* 15(6):669–670.
48. Pinn, V.W. 2007. Introduction: Fourth annual NIH interdisciplinary women's health research symposium November 15, 2007. *Journal of Women's Health* 16(8):1093–1094.
49. Pinn, V.W. 2008. Introduction: Fifth annual NIH interdisciplinary women's health research symposium November 19, 2008. *Journal of Women's Health* 17(8):1227–1228.
50. Pinn, V.W. 2009. Introduction: Sixth annual NIH interdisciplinary women's Health research symposium November 17, 2009. *Journal of Women's Health* 18(10):1487–1488.
51. Pinn, V.W. Abstracts from the NIH Office of Research on Women's Health. Second Annual Interdisciplinary Women's Health Research Symposium: October 20, 2005. *Journal of Women's Health* 15(6):671–691.
52. Pinn, V.W. Abstracts from the NIH Office of Research on Women's Health. Fourth Annual Interdisciplinary Women's Health Research Symposium: November 15, 2007. *Journal of Women's Health* 16(8):1095–1115.
53. Pinn, V.W. Abstracts from the NIH Office of Research on Women's Health. Fifth Annual Interdisciplinary Women's Health Research Symposium: November 19, 2008. *Journal of Women's Health* 17(8):1229–1255.
54. Pinn, V.W. Abstracts from the NIH Office of Research on Women's Health. Sixth Annual Interdisciplinary Women's Health Research Symposium: November 17, 2009. *Journal of Women's Health* 18(10):1489–1523.
55. Hultgren, S., and J. Elam. 2008. Infectious disease and women's health. *Nature Reviews Microbiology* 6(4):254.
56. Domino, S.E., Y.R. Smith, and T.R.B. Johnson. 2007. Opportunities and challenges of interdisciplinary research career development: Implementation of a women's health research training program. *Journal of Women's Health* 16:256–261.
57. Center for Scientific Review. http://www.csr.nih.gov/welcome/orgchart_new.asp. Accessed 28 Jan 2010.
58. The Jacobs Institute of Women's Health and The George Washington University School of Public Health and Health Services. 2009. *Women's health and health care reform: The economic burden of disease in women*. Washington, DC: Jacobs Institute of Women's Health.
59. Weisman, C.S. 2000. The trends in health care delivery for women: Challenges for medical education. *Academic Medicine* 75(11):1107–1113.
60. Weisman, C.S., and G.L. Squires. 2000. Women's health centers: Are the national centers of excellence in Women's health a new model? *Women's Health Issues* 10(5):248–255.
61. Henrich, J.B., J.T. Chambers, and J.L. Steiner. 2003. Development of an interdisciplinary women's health training model. *Academic Medicine* 78(9):877–894.
62. Nicolette, J., and M.B. Jacobs. 2000. Integration of women's health into an internal medicine core curriculum for medical students. *Academic Medicine* 75:1061–1065.

63. Agency for Healthcare Research and Quality, USPSTF. 2009. *The guide to clinical preventive services, 2009: Recommendations of the U.S.* Preventive Services Task Force. Rockville, MD.
64. American College of Obstetricians and Gynecologists. 2007. *Guideline for women's health care: A resource manual*, 3rd edn. Washington, DC: ACOG.
65. Flexner, A., and Pritchett, H.S., 1910. *Medical education in the United States and Canada.* New York, NY: Carnegie Foundation for the Advancement of Teaching, 178–179.
66. Leadley, J. 2009. *Women in U.S. academic medicine: Statistics and benchmarking, 2008–2009*. Washington, DC: American Association of Medical Colleges.
67. Hill, C., Corbett, C., and St. Rose A. 2010. *Why so few? women in science, technology, engineering, and mathematics*. Washington, DC:AAUW.
68. Andrews, N.C. 2007. Climbing through medicine's glass ceiling. *NEJM* 357:1887–1889.
69. Handelsman, J., N. Cantor, M. Carnes, D. Denton, E. Fine et al. 2005. More women in science. *Science* 309(5738) 1190–1191.
70. Leboy, P. 2008. Fixing the leaky pipeline: Why aren't there many women in the top spots in academia? *The Scientist* 22(1):67.
71. Office of Research on Women's Health, National Institutes of Health. 1995. *Women in biomedical careers: Dynamics of change*. Bethesda, MD: NIH, DHHS.
72. Hunt, J. 2010. Why Do Women Leave Science and Engineering? National Bureau of Economic Research Working Paper 15853.
73. National Academies. 2007. *Beyond bias and barriers: Fulfilling the potential of women in academic science*. Washington DC: The National Academies Press.
74. NIH/HHS News Release, January 29, 2007. *NIH leads effort to help women in science and medicine fulfill potential,* 29 Jan 2007.
75. http://womeninscience.nih.gov/index.asp. Accessed 1 Apr 2010.
76. Pinn, V.W., and J.Y. Kravitz 2010. NIH efforts to promote and sustain the careers of women in science. *ASCB Newsletter*. 13–14, Mar 2010.
77. Research on Causal Factors and Interventions that Promote and Support the Careers of Women in Biomedical and Behavioral Science and Engineering RFA GM-09-012. NIH Guide. 2008. http://grants.nih.gov/grants/guide/rfa-files/RFA-GM-09-012.html. Accessed 1 Apr 2010, July 14, 2008.
78. NIH/HHS News Release 2009. *NIH awards grants to examine factors influencing women's careers in science*. 9 Oct 2009.
79. NIH Office of Extramural Research. Frequently Asked Questions Policies Related to Parental Leave and Child Care. 2010. http://grants.nih.gov/training/faq_childcare.htm. Accessed 1 Apr 2010.
80. Mid-Atlantic Higher Education Recruitment Consortium. http://www.midatlanticherc.org/home/index.cfm?site_id=4076. Accessed 1 Apr 2010.
81. NIH Office of Extramural Research. Research portfolio online reporting tool. research grants: Average size, by gender. http://www.report.nih.gov/nihdatabook/Default.aspx#15. Accessed 1 Apr, 2010.
82. Office of Research on Women's Health, National Institutes of Health. 2009. *Women in biomedical research: Best practices for sustaining career success*. Bethesda, MD: NIH, DHHS.

Chapter 5
Part-Time Faculty and Their Hidden Value

Julia V. Johnson and Erin E. Tracy

Abstract A desire for improved worklife balance has prompted more faculty to work less than one full-time equivalent. While not exclusive, women faculty are especially to make this request. This chapter explores the hidden value of part-time faculty and the perceptions of department chairs and their department managers about these special faculty members.

Keywords Part-time practice · Barriers · Productivity

The success of an academic department is dependent on the recruitment and retention of dedicated clinicians, educators, and researchers. To attract the best faculty, institutional leaders must recognize changes in the workforce and adjust the work environment appropriately for the next generation of academicians. It has been recognized that new faculty are increasingly interested in work-life balance; a major change is the request for part-time positions. The desire for less than full time hours was identified over 25 years ago [1], but medical schools now recognize that current and future faculty will negotiate for part-time positions [2]. This has raised some concerns for institutional leaders [3], although data suggests that part time faculty make excellent contributions to the department and institution [4]. It is critical for all academic institutions to embrace this change in the work culture to maintain the best faculty at our medical schools. This chapter will review the history of part-time faculty positions in academic medical centers, consider the current and future incidence of part-time faculty, evaluate productivity and patient care provided by part-time faculty, review the impact of part-time hours on the faculty career success, and consider the future of part-time faculty positions at academic institutions. It is the objective of this chapter to review the evolution of part-time academic faculty positions and develop strategies to optimize the career development part-time faculty.

J.V. Johnson (✉)
Department of Obstetrics and Gynecology, University of Massachusetts Memorial, Worcester, MA 01605-2903, USA
e-mail: Julia.johnson@umassmemorial.org

W.F. Rayburn, J. Schulkin (eds.), *Changing Landscape of Academic Women's Health Care in the United States*, International Library of Ethics, Law, and the New Medicine 48, DOI 10.1007/978-94-007-0931-7_5,

5.1 History of Part-Time Faculty Positions

There is an increase in the number of part-time workers throughout the United States. A recent survey demonstrated that more than 10% of professionals elect to work part-time [5]. The American Association of University Professors recognized the need to address part-time faculty positions in 1980 and assessed policies regarding the academic freedom and tenure options for part-time faculty. Work hours for medical professionals; however, has not been well studied until very recently. The earliest discussion of part-time faculty positions can be found in the nursing profession. In 1970, the first study examining the job satisfaction and educational effectiveness of part-time faculty demonstrated that nursing faculty remained in their positions longer and had greater job satisfaction when allowed to decrease hours [6]. In addition, the students rated these faculty highly and did not perceive an effect on the quality of their education. Although the length of time to advancement is increased for part-time faculty, the overall job satisfaction and ultimate career development is equivalent to full-time nursing faculty. More recent studies demonstrate that part-time faculty actually are ranked highly by students than full-time faculty and are considered to be superior educators [7]. Although the challenge of decreasing work hours is recognized, nursing programs have embraced the use of part-time faculty.

The interest in part-time faculty positions for physicians was not identified until 1989, when a study examined the impact of family on career development [8]. Women faculty who held full-time faculty positions in departments of medicine answered a survey that examined the effect of childbearing and child rearing on professional development. Although the majority of faculty surveyed were pleased with their family decisions, seventy-eight reported that this slowed or markedly slowed their career progress. This was the first evidence that work-life balance was challenging for full-time faculty who wished to advance in their careers.

In 1993, the Association of American Medical Colleges (AAMC) published the first study that examined the increase in part-time faculty positions [1]. Sixty-three percent of the part-time faculty surveyed were men, and 86% of the part-time faculty were on a non-tenure tract. Interestingly, half of the part-time faculty stated that they developed the positions themselves; very few medical schools had policies that addressed part-time positions. The career limitations of part-time faculty were clear in this study. Only 8% reported that their institution would allow additional time to accomplish tenure or promotion for part-time faculty. And both men and women part-time faculty reported that there were false perceptions and lack of respect from their full-time colleagues. However, overall the faculty were pleased with their positions; overall career satisfaction was rated 8.6/10 for part-time faculty.

Since the AAMC study, there has been an increase in the number of part-time positions. In addition to the increase in requests for these positions, there is evidence of improved opportunities for young faculty who request a less than full-time position. The desire for work-life balance is believed to be related to the increase in women physicians, two physician families, and the generational change in the work force. The "Generation X" faculty express concern about negative impact of work on family and are less likely to put their jobs ahead of family and friends [9]. It is the impression of younger faculty that their parents "loyalty" to their job did not lessen the risk of job loss or increase job satisfaction. These new faculty will not value their academic institution above their career and will not hesitate to change jobs if work-life balance is not valued by those in authority. There is an increased desire for part-time professional positions. There is strong evidence in the nursing profession that part-time faculty are productive and have increased job satisfaction. Although medical faculty fall behind other professionals in seeking work-life balance, it is clear that this will be a common request for junior faculty. Academic leaders must recognize that faculty recruitment and retention will be adversely effected if part-time positions are not a career option for their faculty.

5.2 Incidence of Part-Time Faculty

Despite the limited early data, it is evident that the number of part-time academic faculty positions is increasing. An American Medical Association (AMA) survey in 2001 demonstrated that there are up to 25% of the faculty are part-time (Fig. 5.1). In women's health care specialties, such as Obstetrics and Gynecology, Family Medicine, and Internal Medicine, the percentage of surveyed faculty that had ever worked part-time was 20, 18, and 16% respectively. This is compared to 13% of other faculty surveyed. This demonstrates that women's health care physicians have a strong interest in part-time faculty positions.

Studies in the field of Pediatrics, Radiology, Internal Medicine, and Obstetrics and Gynecology demonstrate the increased number of part-time faculty positions. In Pediatrics, one study reported an increase in part-time positions from 11% in 1993 to 15% in 2000 [10]. A resident survey demonstrated that 40% of Pediatric residents were seeking a part-time position, although only 20% had successfully been recruited to a part-time positions [11].

In areas of women's health care, such as Internal Medicine, the number of part-time faculty increased from 9.9% in 1996 to 29.2% of faculty in 2004 [12]. A recent survey of Chairs in Pediatrics, Internal Medicine, Family Medicine, and Surgery demonstrated that the percentage of departments with part-time faculty was 94,

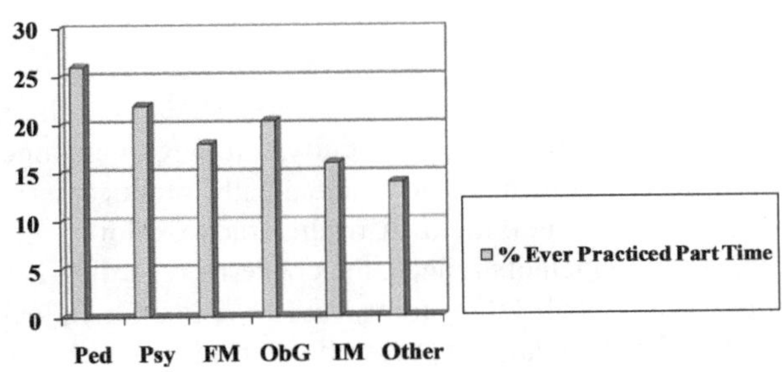

Fig. 5.1 Patient care physician survey (AMA, 2001; E Tracy)

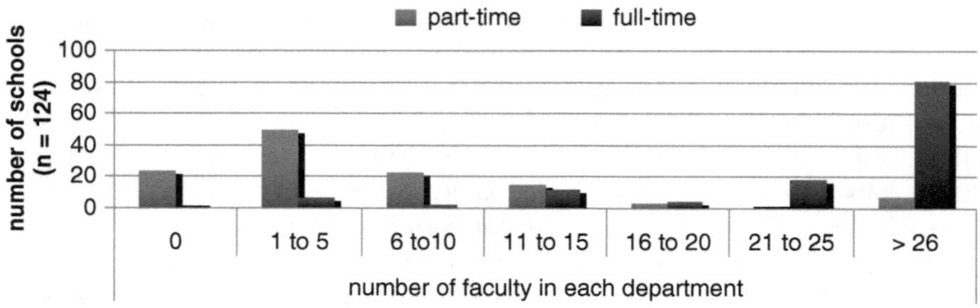

Fig. 5.2 Number of part-time and full-time faculty

89, 86, and 72% respectively [13]. A recent study demonstrated that 17% of academic Obstetrician Gynecologists were part-time faculty [14]. Eighty-two percent of departments had part-time faculty, consistent with other departments that provide women's health care. The number of part-time faculty in the OB/GYN departments were most commonly between 1 and 5; although six departments reported more than 15 part-time faculty (Fig. 5.2). Interestingly the number of men ($n = 445$) and women ($n = 539$) was similar, although a greater percentage of women than men worked part-time (23% vs. 16%). The Women's Physician Congress of the AMA conducted a survey in 2008 (Unpublished data, E Tracy) demonstrated that the primary reason for part-time work is for raising children (68%), followed by achieving work-life balance. It is critical to recall, however, that only 30% of women in this survey had ever worked part-time; 70% had always been full-time physicians.

There is a significant increase in part-time faculty in all academic departments, with the greatest increase in departments that provide care to women, such as OB/GYN, Family Medicine, and Internal Medicine. The OB/GYN survey found that over 90% of Chairs predicted an increase in part-time faculty in the future. It is critical for academic leaders to consider this change in the work force to assure that new faculty can be recruited into academic departments.

5.3 Productivity and Quality of Patient Care for Part-Time Faculty

Academic leaders have raised concerns about both the productivity and the quality of care offered by part-time faculty. Although part-time nursing faculty have been shown to be clinically and academically productive, studies examining part-time medical faculty is limited. A recent study exploring the attitudes of Internists and their Division Chiefs reflects the concerns regarding academic success and clinical productivity [15]. Interestingly, the Division Chiefs expressed more benefits than drawbacks from part-time work by their faculty (see List 5.1). Some of the benefits includes the ability of the leader to recruit and retain high quality faculty, and allow a more diverse workforce. The challenges included managing work and pay equity and assuring scholarly productivity.

Recent studies have begun to reassure academic leaders and faculty of the value of part-time faculty. Chairs in Pediatrics, Internal Medicine, Family Medicine, and Surgery rated their experiences as very and extremely satisfied with their part-time faculty [13]. One respondent noted that part-time faculty: "allow inclusion of talented, capable individuals who have other commitments (i.e. family)". Interestingly, the Chairs noted that part-time faculty "work more than their allocated time". The clinical productivity was seen to exceed expectations. Sixty-two percent of the Chairs said they were very satisfied with part-time faculty; only 7% we dissatisfied.

A study of part-time academic radiologists demonstrated that part-time faculty expected to work one-half time, worked 32.3 h/week compared to 51.2 h/week for full-time faculty [16]. For primary care physicians, the relative value units (RVUs) was significantly greater for part-time faculty compared to full-time faculty (2.1 RVUs/hr vs. 1.3 RVUs/hr; $p < 0.01$) [17].

List 5.1 Division chiefs' views of positive and negative consequences of part-time work for physicians

Benefits of having part-time faculty in divisions

- Recruitment and retention of high quality faculty members
- Increases unpaid work for division
- Increases division visibility (i.e. time to present at national meetings)
- Promotes diversity of workforce
- Creates role models for work-family balance
- Creates new career paths
- Breaks down cultural constraints of full-time work
- No benefits were identified by some division chiefs

Challenges of having part-time faculty in division

- Assuring equal pay among division members
- Assuring equality of work among division members
- Defining number of hours as to what is full-time and part-time
- Defining and providing health insurance and benefits
- Monitoring motivation for work
- Determining expectations for scholarly productivity

Negative consequences of part-time work for worker

- Lower pay
- Fewer or no benefits and heaith insurance
- Slower academic career trajectory

- Less protected time per FTE
- Potential to be viewed as less committed worker
- Not fully integrated into the division

Part-time faculty express concerns about the quality of patient care they offer when working less than full-time [4]. However, the quality of clinical care provided by part-time faculty appears to be equivalent to or greater than that provided by full-time faculty. Using qualitative measures in an ambulatory setting, such as cancer screening and diabetes management, part-time faculty were found to offer a higher standard of care compared to their full-time colleagues [17]. Patient satisfaction was also shown to be high for part-time faculty [18]. Compared to full-time faculty, there was no difference in patient trust and patient satisfaction. Patients expressed no concern regarding their physician's part-time hours.

Although the data is limited, it appears that both productive and the quality of patient care is maintained when faculty work part-time hours [19]. Both part-time physicians and their leaders express concern regarding the effect of the decreased hours on academic practice, the satisfaction of Division Chiefs and Chairs, as well as the patients care for by part-time faculty, are high. There is no evidence that part-time hours adversely effects clinical care.

5.4 Career Development for Part-Time Faculty

It is clear that part-time faculty are excellent health care providers and bring productivity to academic departments. However, there is a potential adverse effect on their career. The ability to advance through the tenure tract and achieve promotion and leadership positions appears to be limited by working part-time. Indeed, Division Chiefs express concerns regarding "slower academic career trajectory" for the part-time faculty [15].

Part-time faculty also identify these drawbacks. They feel "marginalized" and identify that "Institutional culture does not support part-time/lack of leadership support"(see List 5.2). To assure that career advancement is possible for part-time faculty, it is critical for institutional policies to address the needs of this increasing group of academic faculty. Policies that extend time to achieve tenure, increase research support, and assure adequate mentoring are needed to assure retention of part-time faculty.

Unfortunately, a review of policies for the "top ten" academic medical schools demonstrated that such policies are rare [2]. In 2008, only three of these academic institutions reported a part-time faculty policy, two had limited support mechanisms for part-time faculty, and two would only allow part-time positions under extreme circumstances. Another survey demonstrated that only 25 of 95 medical schools with a tenure systems allowed part-time faculty to seek tenure [20]. Interestingly, extending the tenure clock was allowed for a multitude of reasons, but only 23 schools allowed part-time hours as a rationale for extending the time to tenure.

List 5.2 Positive and negative consequences of part-time work for faculty participants

Career benefits

- More research time
- Enabled worker to focus on career goals
- Worker as role model to others
- Transition from clinician educator to clinician investigator
- Able to pursue new scholarship
- Able to pursue advance degree

Personal benefits

- Greater quantity and quality of family time
- Ability to participate in children's activities, mom's groups
- Increased time for school and community involvement
- Time for journaling/creative writing
- Time for spirituality
- Time for exercise and self-care

Drawbacks

- Less pay and loss of benefits
- More work than FTE
- Slow to promotion/overlooked for career opportunities
- Demotion
- Less desirable work
- Institutional culture not supporting part-time/lack of leadership support
- Lack of support from colleagues
- Being marginalized within the division

Fortunately, there has been a recent effort to develop academic policies for part-time faculty. The Federation of Pediatric Organizations Task Force, describes policies that optimize retention of valued part-time faculty [21]. Options described by this document include structured mentoring programs, extending time to tenure, decreasing teaching and clinical duties to allow time for research, offering child care support, allow flexibility with part-time positions, and increasing laboratory/research support. The Task Forces speaks of replacing a culture of "hierarchical shame" with "horizontal supportive" environment. Although the AAMC noted that faculty appointment and tenure track have improved for part-time faculty over the past 25 years [22], there continue to be significant barriers. Even institutions with policies to support part-time faculty may discourage these policies or allow faculty ignorance of their options.

Another significant concern for part-time faculty is financial issues. Although faculty accept decreased salary when decreasing their hours, there is a concern with decreased or lost benefits. A 2008 survey by the Women's Physician Congress of the AMA demonstrated the greatest barrier to part-time work being loss of benefits.

5.5 2008 WPC Survey Barriers to Part Time Work (Data from E Tracy)

- Loss of benefits (61% – "very significant barrier")
- Lack of part time opportunities (61%)
- Loss of income (58%)

A recent survey of medical schools revealed that while the majority of part-time faculty are offered retirement benefits, as well as health, life, disability, and dental insurance, the cost to the faculty member is high [23]. And many professional liability carriers do not adjust their rates based on the faculty's part-time status, raising the cost to the department. Fortunately the AMA has compiled a list of carriers that offer reduced premiums to part-time physicians.

Clearly, there are limitations for part-time faculty. In addition to challenges in academic advancement, the cost of their part-time status may be prohibitive. It is critical for academic institutions to develop and execute policies that support part-time faculty to assure that they will continue to be valuable members of the faculty.

5.6 The Future

A recent report from a task force, convened in 2007 by the Associate of Specialty Professors, came to the agreement that part-time faculty can "enrich a department of medicine, enhance workforce flexibility, and provide high quality research, patient care, and education in a cost–effective manner" [24]. This task force provides valuable steps to optimize part-time practice including addressing fixed costs, malpractice costs, space, cross-coverage, mentoring, career development, productivity, and scheduling. The costs of part-time work considered quantitative factors and qualitative factors, such as the mission of the academic institution, workplace culture, and physician morale. The part-time principles from the Medical College of Wisconsin, led by Dr. Ann Nattinger, was provided as the standard for academic leaders when developing policies regarding part-time faculty:

1. Be *nonjudgmental* about who is going part-time and why; do not assume that part-time is always for young women with children.
2. Create *transparent and fair* policies; avoid jealousy from full-time faculty.
3. *Standardize* processes for urgent clinical matters so part-time faculty participate equitably.

4. Provide *flexibility* for the part-time faculty as much as possible.
5. Consider providing *full (not prorated) professional development* dollars to all faculty regardless of FTE.

It is critical for leaders to accept the changes in the work force which have led to an increased desire for part-time positions. For academic departments to succeed leaders need to increase programs to support work-life balance, educate all faculty regarding the document quality of part-time faculty, and prevent negative impressions by full-time faculty on the academic value of their part-time colleagues.

5.7 Conclusions

Part-time faculty are increasing in number and importance in academic departments. The recruitment and retention of excellent faculty requires that leaders recognize the increased desire of faculty for work-life balance. The history of part-time faculty in women's health care specialties is limited, but the available data suggests that these faculty are highly productive and offer high quality patient care. Indeed, studies suggest that departmental leaders, Chairs and Division Chiefs, appreciate the dedication of their part-time faculty. These faculty, although they appreciate the opportunity to work less than full time, do express concerns regarding academic advancement and financial limitations. It is critical for academic leaders to develop effective policies to assure that part-time faculty are able to meet their potential. Academic institutions must encourage and support programs that assure the clinical, educational, and research success of their part-time faculty.

References

1. Levinson, W., K. Kaufman, and J. Bickel. 1993. Part-time faculty in academic medicine: Present status and future challenges. *Annals of Internal Medicine* 119:220–225.
2. Bristol, M.N., S. Abbhul, A.R. Cappola, and S.S. Sonnad. 2008. Work-life policies for faculty at the top ten medical schools. *Journal of Women's Health* 17:1311–1320.
3. Socolar, R.R.S., and L.S. Kelman. 2002. Par-time faculty in academic pediatrics, medicine, family medicine, and surgery: The views of the chairs. *Ambulatory Pediatrics* 2:406–413.
4. Kahn, J.A., S.J.F. Degen, M.E. Mansour, E. Goodman, M.H. Zeller, T. Laor, N.E. Lanphear, and T.F. Boar. 2005. Pediatric faculty members' attitudes about part-time faculty positions and policies to support part-time faculty: A study at one medical center. *Academic Medicine* 80:931–939.
5. Corwin, V. 2001. Five strategies of successful part-time work. *Harvard Business Review* 79:121–127.
6. McPhall, F.J. 1979. Use of part-time teachers benefits students and faculty. *Canada Nurse* 66:36–37.
7. Williams, P.L. 1994. A comparison between the roles of part-time and full-time clinical tutors: Is there a difference? *Nurse Education* 14:427–435.
8. Levinson, W., S.W. Tolle, and C. Lewis. 1989. Women in academic medicine: Combining career and family. *New England Journal of Medicine* 321:1511–1517.

9. Bickel, J., and A.J. Brown. 2005. Generation X: Implications for faculty recruitment and development in academic health center. *Academic Medicine* 80:205–210.
10. Cull, W.L., H.J. Mulvey, O'Connor KG, D.R. Sowell, C.D. Berkowitz, and C.V. Britton. 2002. Pediatricians working part-time: Past, present, and future. *Pediatrics* 109:1015–1020.
11. Cull, W.L., G.L. Caspary, and L.M. Olson. 2008. May pediatric residents seek and obtain part-time positions. *Pediatrics* 121:276–281.
12. Lugtenberg, M., P.M. Heiligers, J.D. deJond, and L. Hingstman. 2006. Internal medicine specialists attitudes towards working part-time. *BMC Health Services Research* 6:126–135.
13. Socolar, R.R., and L.S. Kelman. 2002. Part-time faculty in academic pediatrics, medicine, family medicine, and surgery: The views of the chairs. *Ambulatory Pediatrics* 2:406–413.
14. Rayburn, W.F., B.L. Anderson, J.V. Johnson, M.A. McReynolds, and J. Schulkin. 2010. Trends in the academic workforce in obstetrics and gynecology. *Obstetrics and Gynecology* 115:141–146.
15. Harrison, R.A., and J.L. Gregg. 2009. A time for change: An exploration of attitudes towards part-time work in academia among women internists and their division chiefs. *Academic Medicine* 84:80–86.
16. Cheroff, J.D., C.E. Bird, and B.C. Amick. 2001. Career paths in diagnostic radiology: Scope and effect of part-time work. *Radiology* 221:485–494.
17. Fairchild, D.G., Sax-McLoughlin, K., S. Gharib, J. Jorsky, M. Portnow, J. Richter, N. Gagliano, and D.W. Bates. 2001. Productivity, quality, and patient satisfaction: Comparison of part-time and full-time primary care physicians. *Journal of General Internal Medicine* 16:663–667.
18. Mechaber, H.F., R.B. Levine, L.B. Manwell, M.P. Mundt, and M. Linzer. 2008. Part-time physicians: Prevalent, connected, and satisfied. *Journal of General Internal Medicine* 23:300–303.
19. Parkerton, P., E. Wagner, D.G. Smith, and H.L. Straley. 2003. Effect of part-time practice on patient outcomes. *Journal of General Internal Medicine* 18:717–724.
20. Socolar, R., L. Kelman, C. Lannon, and J. Lohr. 2000. Institutional policies of U.S. medical schools regarding tenure, promotion, and benefits for part-time faculty. *Academic Medicine* 75:846–849.
21. Alexander, D., T. Boat, M. Britto, A. Burke, A. Homes, T. Sectish, B. Atnton, and D. Stevenson. 2009. Federation of pediatrics organization task force on women in pediatrics: Consideration for part-time training and employment for research-intensive fellows and faculty. *Journal of Pediatrics* 154:1–3.
22. Bunton, S.A., and W.T. Mallon. 2007. The continued evolution of faculty appointment and tenure policies at U.S. medical school. *Academic Medicine* 82:281–289.
23. Socolar, R.R., L.S. Kelman, C.M. Lannon, and J.A. Lohr. 2000. Institutional policies of U.S. medical schools regarding tenure, promotion, and benefits for part-time faculty. *Academic Medicine* 75:846–849.
24. Linzer, M., C. Warde, R.W. Alexander et al. 2009. A report from the Association of Specialty Professor part-time careers task force on behalf of the Alliance for Academic Internal Medicine. *Academic Medicine* 84:1395–1400.

Chapter 6
Trends in Faculty Salaries

William F. Rayburn

Abstract A concern in women's health care is the potential for inequitable compensation for services provided. For example, a 2000 JAMA report indicated that inflation-adjusted annualized salary growth was negative for faculty in obstetrics and gynecology. This chapter examines trends in the salaries of medical school faculty in women's healthcare based on a review of extensive national salary databases.

Keywords Salaries · Faculty · Income

6.1 Introduction

Salaries of academic faculty do not usually rival those of physicians of the same specialty in the community. Nevertheless, faculty enjoy a varied professional lifestyle which encompasses teaching and scholarly activity, along with patient care. A growing faculty size provides a relatively large group practice, so that in-hospital call is often less onerous [1].

Economic changes in the 1990s markedly affected protected time for academic duties and, in some cases, widened the differences in salaries between private practice and academic physicians. The "80 h work week" and caps on the numbers of residents in training programs from a provision of the Balanced Budget Act of 1997 further increased the clinical demand on faculty and increased the number of residents beyond the allowable payment cap [2].

Personal, departmental, and institutional needs in this fiscal environment created pressure to increase clinical productivity. Reimbursement cuts to health care institutions resulted in tremendous pressure to fill beds and generate ancillary revenues to remain solvent. As a result, the proportion of clinically derived monies to total

---

W.F. Rayburn (✉)
Department of Obstetrics and Gynecology, University of New Mexico, Albuquerque, NM 87131, USA
e-mail: wrayburn@salud.unm.edu

W.F. Rayburn, J. Schulkin (eds.), *Changing Landscape of Academic Women's Health Care in the United States*, International Library of Ethics, Law, and the New Medicine 48,
DOI 10.1007/978-94-007-0931-7_6, © Springer Science+Business Media B.V. 2011

medical school revenues grew from approximately 12% in the early1970s to more than 50% in the 2000s [3].

This evolving need for increasing clinical volume altered the balance of these traditional faculty responsibilities. As economic pressures mount and budgets shrink, academic departments are exploring ways of paying faculty to help maintain the global missions of the organization. This chapter will examine trends in the salaries of medical school faculty, particularly those in women's health care, based on reviews of the current literature and an extensive national salary database.

6.2 Fundamental Themes

Every academic department is unique because of the inherent complexities, including its financial infrastructure. The underlying value system emphasizes academic productivity and success for the department as a whole. Not every faculty member is or will be an academic star, so role fulfillment and goal attainment are stressed. The following are fundamental themes when faculty compensation is considered at recruitment and annually.

6.2.1 Transparency and Fairness

As a sociologic group, physicians and faculty strongly desire self-esteem, fairness, and security [4]. Ideally, the entire department faculty should create or agree on a method for compensation. Any system can fail or cause divisiveness. Despite best efforts in being "fair," any system for compensation is subject at any time to complaints from certain faculty. Leadership, trust, and transparency are important factors at any workplace environment. In particular, division directors should be aware of specifics in the compensation system and how to implement any incentive or bonus portions.

6.2.2 Culture and Leadership

Faculty salaries are not a sole predictor of a department's success. The stronger the culture for academic productivity, the less need for financial incentives to foster scholarly and educational efforts. Creating the appropriate culture for productivity is complex and requires constant vigilance. A culture of financial and scholarly productivity is linked to department leadership and the hiring of faculty members with strong internal incentives to be successful.

6.2.3 Faculty Activity Database

Every faculty member should pay attention to his or her accomplishments and goals achieved during the current academic year. Completing forms by each member either annually or semi-annually is now customary to document their research, teaching, and service activities. The chair uses this information in an annual review with each faculty member. Decisions about rewarding academic productivity are made at least partially on the basis of this activity database.

6.2.4 Work RVUs

A myth is that the academic clinician does not work as hard as a community physician. Information has been reported by the Medical Group Management Association (MGMA) about "work" output, as measured by relative value units (RVU) developed by the Centers for Medicare and Medicaid Services (CMS) [5]. The work effort by the academic physician was found to equal or exceed that of counterparts in private practice. Although work RVUs may not be the best measure of clinical "work," it is a mainstay for comparison with others in the same specialty at other medical schools.

6.2.5 Annual Bonuses

Although academic success is a lofty goal of all faculty, chairs are well grounded in the reality that clinical revenues drive the department. All departments and centers derive most, if not all, of their bonus pool funds from clinical collections. A common need of chairs is to have some form of a performance-based annual bonus as a relatively small percentage (e.g. 10% or less) of overall compensation.

6.2.6 Discretionary Funds

In addition to annual bonuses, most chairs need some flexibility in distributing department funds to faculty. This could open the door for criticism from some by faculty about favoritism or unfairness, yet it does provide discretionary funds with some "wiggle room" to recognize other's contributions that may not fit neatly into any existing compensation formula.

6.2.7 Tenure

Although academic physicians have historically received salaries that are lower than their peers in private practice, many are provided tenure with its job security, to make it attractive to remain at their academic institution. Reasons for the best and brightest graduates to consider a position in academia include a reasonable clinical workload, research/scholarly opportunities, attainment of tenure, and college tuitions of dependent children at certain centers. However, as the market changes with health care reform, each of these reasons including tenure will be more scrutinized. Many academic health centers are now faced with the difficult task of restructuring faculty tenure policies to improve the bottom line, while risking disincentives to continue an academic career.

6.3 Trends in Faculty Compensation

A long-standing primary source for salaries according to specialty, rank, type of school (private or public), and region is the *Report on Medical School Faculty Salaries*, published annually be the Association of American Medical Colleges (www.aamc.org). Salaries are reported at the 25th, median, 75th percentile, and mean. The compensation data are accumulated each October from the preceding academic year from each medical school. The total compensation includes the contractual salary and supplemental earnings (medical practice supplement and bonus/incentive pay). Uncontrolled outside earnings are reported by only a minority of faculty. Other influences on salaries such as fringe benefits, retirement contributions, or tuition remissions for family are not included in the calculations.

6.3.1 The 1990s

Clinical revenue from faculty practice plans is important but especially critical to smaller clinical departments, such as obstetrics and gynecology, which depend more on patient care than on research dollars. Since the 1990s, intensified price competition by payers has led to a flattening or decline in reimbursements from both public and private payers [6]. Any reduced reimbursement per unit of service can be detrimental to faculty recruitment and retention and to certain mission-based activities with fixed costs such as education and research. In light of changes in patient care reimbursement, growth in the salaries of academic physicians requires close monitoring.

While actual average medical school faculty salaries increased in the 1990s, the growth rate of average clinical faculty salaries increased less or declined when adjusting for inflation. A report by the Association of American Medical Colleges (AAMC) found that overall growth rates of most clinical faculty salaries increased slightly above inflation between 1988 and 1998 [7]. Those findings are shown in Fig. 6.1 in which growth rates are based on mean salaries in constant (1988) dollars, with 1993 and 1998 means weighted to match the faculty rank distribution in 1988. The proportionate change in real mean salary (base year, 1988) related to department and faculty rank. Inflation-adjusted annualized compound growth of clinical faculty salaries declined from 1.9% per year (1988–1993) to 0.2% per year (1993–1998). From 1993 to 1998, inflation-adjusted salary growth rates were negative for obstetrics and gynecology (–0.5%) or positive for family practice (+2.7%) and internal medicine (+0.8%). Significant differences in salary growth also related to school characteristics (e.g., geographic region, public vs. private, community-based vs. non-community-based, and research intensity).

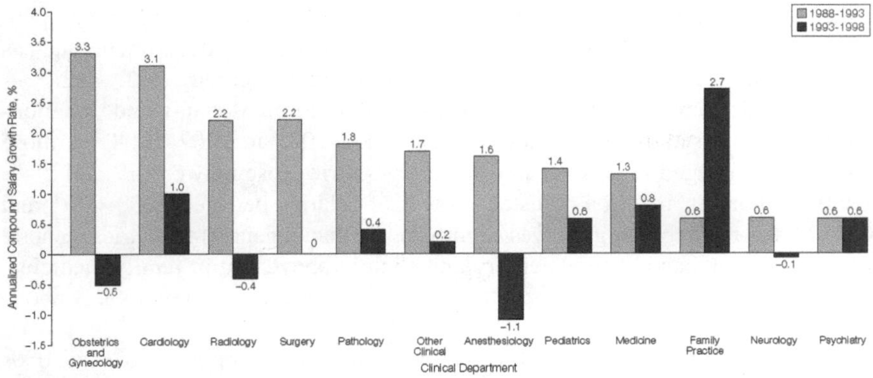

Fig. 6.1 Annualized inflation-adjusted compound growth rates of clinical faculty salaries by department, 1988–1998 (from [7] with permission). Growth rates are based on mean salaries in constant (1988) dollars, with 1993 and 1998 means weighted to match the faculty rank distribution observed in 1988

6.3.2 The 2000s

Overall faculty salaries increased at a rate comparable to the inflation for the period 1988–1998 [8]. Noted exceptions were compounded salary increases in family medicine and decreases in ob-gyn. A more recent review of the AAMC annual Faculty Salary Survey revealed changes in compensation for general faculty in

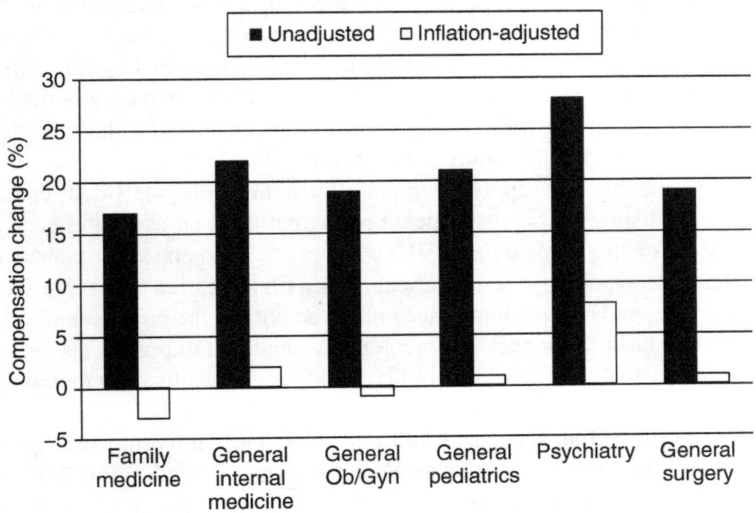

Fig. 6.2 Average annual change in inflation-adjusted compensation from academic years 2000–2001 to 2008–2009 for general faculty in core teaching clinical specialties (* $p < 0.05$ compared with cumulative rate of inflation using Bonferroni correction (from [9] with permission)

women's health care specialties that were very similar [9]. Changes in compensation were comparable for generalists in core clinical teaching departments such as internal medicine, family medicine, and obstetrics and gynecology.

Figure 6.2 compares changes in unadjusted (current dollars) and inflation-adjusted compensation from academic year 2000–2001 to 2007–2008 for those departments engaged in women's health care. Salaries rose between 17 and 28%. Equally important, inflation-adjusted growth of salaries between 2000–2001 and 2007–2008 for general ob-gyns was slightly below that in general pediatrics, general internal medicine, and general surgery and slightly above that in family medicine. Only faculty in psychiatry saw median salary increases that were remarkably above the inflation rate.

Downward trends in patient care compensation were reported in the 2000s and may result from changes in reimbursement, more competition, fewer surgeries, and more profitable assisted reproductive technology moving to the private sector [6, 10].

6.4 Aligning Compensation with Education

It is difficult to place a financial value on teaching. There are limitations in resources related to decreased government funding for graduate medical education and increasing demands on faculty time, along with increased demand for public accountability. Increasingly, leaders of academic medical centers have recognized the importance of developing systems that specifically assign resources in support of all academic missions, but especially for educating students and residents [11]. In 1999, Watson and Romrell reported a process that came to be known as "mission-based budgeting (MBM)" [12]. The three-step process described by the University of Florida group consisted of identifying revenue streams to fund each of the institution's missions, evaluating each faculty member's productivity with regard to each mission, and aligning funding source with faculty effort [13].

Many departments now apply a template for approaching MBM in education, beginning with listing all faculty educational activities, then assigning each activity a weight in relative value units (RVUs). Factors recommended for consideration included the time to perform the educational tasks, time required to prepare a level of faculty expertise, and relative importance of the activity to the professional development of the institution's trainees. A potentially difficult but important task is linking compensation to quality of teaching, rather than focusing exclusively on quantity of work [12].

Funding for most clinical department's educational mission originates from three sources: state appropriations, Medicare Direct Medical Education (DME) funding, and tuition. Distribution of departmental funds, including the portions of clinical revenues and state-allocated funds, was the prerogative of the department chair in the past. Except for education administration, no clear metric linking these funds with mission is often used for individual faculty.

Table 6.1 Core educational value unit (EVU) allotment of administrative positions in medical education programs, University of New Mexico School of Medicine

Fellowship director	0.10
Residency program director	0.30
Residency associate program director	0.10
Student clerkship director	0.20
Student associate clerkship director	0.10

An Educational Value Unit (EVU) is defined as teaching time spent educating students and residents that is not associated with billable clinical activity. Core EVU time was also allotted for the administration of education, for residency program directors, fellowship program directors, and clerkship directors. The EVU allotment for program administration at my school (as shown in Table 6.1) was taken from national certifying bodies as well as from our own experience.

The alignment of educational expectations with compensation and accountability is essential in any academic department dedicated to women's health care. The metric should be easily understood, quickly implemented, and perceived to be fair by the faculty. After initially attempting to adapt published RVU-based systems to our department's needs, we found the task of enumerating and assigning EVUs to each educational activity as important as assigning a billing code to a clinical activity.

6.5 Performance-Based Compensation for Research

Academic departments must compete for compensation in a dynamic marketplace to recruit and retain the best faculty. Grants and contracts combined with clinical revenues have been the largest sources for faculty support, providing over half of all department funding. Faculty appointments at most schools are placed into several research tracks, each reflecting a different career pathway: (1) physician-scientist track (80% research/20% clinical; tenure) for faculty with major efforts in research and teaching; (2) clinician-educator track (80% clinical/20% research; non-tenure) for faculty with major efforts in clinical service and teaching; (3) basic scientist track (100% research; tenure) for non-physician faculty devoted primarily to research and teaching; and (4) research scientist track (100% research; non tenure) for faculty who primarily support the research mission. The proportion of time commitment and the names of the tracks vary between institutions, but the concept is the same. Most faculty in family medicine, general internal medicine, and obstetrics and gynecology belong to either the physician-scientist or the clinician-educator track.

Tarquinio and colleagues nicely summarized how compensation is performance-based for research at Vanderbilt University Medical Center [14]. Physician-scientists receive an annual base salary and also qualify for a research bonus under certain conditions. Base salaries are usually set at the same level as the prior year plus increase by meeting objective criteria. Major grants are defined as being a minimum 3-year term, involving 15% of faculty effort, and generate funds for recovering indirect costs for the institution.

For physician-scientists who funded their base salaries up to 80% of the NIH's capitated rate, base salaries are increased by an amount determined by the department before each new fiscal year. Physician-scientists who are funded at less than 80% of their base salaries but who submitted a new major grant application during the prior 12 months are often given the institution-recommended increase in base salary [14]. If a physician-scientist did not fund 80% of his or her base salary and did not submit a new major grant application during the past 12 months, any institution-recommended increase would begin at the point during the subsequent fiscal year when a major grant application was submitted.

A bonus for physician-scientist faculty is also possible either for some additional noteworthy research or for reaching some clinical RVU benchmark. The lack of financial incentives to work above the clinical benchmark encourages faculty in the physician-scientist track to maintain their focus on research productivity. Any bonuses, whether for research or clinical duties, do not carry over into subsequent years.

6.6 Salary Contract Negotiation

Each year there are more than 5,000 entry-level faculty who begin their careers in academic medicine. For many, discussing salary and compensation is awkward. In addition, many faculty remain uncertain about salary contract negotiations when switching schools, despite their experience in academia. Several standard steps can be followed to promote agreement and to facilitate the process.

The faculty recruit must decide whether s/he wishes to be offered a full-time or part-time faculty position in the department. Their rank and specialty needs to be understood as well as the probable track to choose. As part of the salary contract, the effort distribution for the first year of appointment is assigned full time equivalents (FTE) for clinical, research, education, and administration tasks. Each year, and as needed, the faculty's effort distribution is re-discussed and the contract modified to reflect any changes in the effort distribution.

A contract salary for the forthcoming fiscal year usually includes a base and a supplement component. In most medical school departments, there is an attempt to augment the contract salary as an incentive component based on meeting expectations of the Accreditation Council for Graduate Medical Education (ACGME) competencies. A one-time incentive may be provided with initiation of the new faculty member's appointment.

In addition to salary, the position carries a variety of fringe benefits. Information about these benefits is customarily provided by a designated staff member in the departments or school's faculty affairs or human resources offices. The recruit's personal expenses in recruitment and moving to the new department are reimbursed according to university policy. The department agrees to pay a reasonable moving package and a trip(s) to the destination city before relocating to identify housing.

Matters such as office and/or laboratory space, secretarial support, assigned duties, and duty location are determined at the discretion of the chair and dean according university rules and regulations.

By 10 years from their initial appointment, about half (48%) of entry-level faculty leave their original department to switch to another school (21%) or leave academia (79%) [15]. In the event that the employment relationship terminates between the medical school and a faculty member, the employee agrees that for a period from the date of termination (2 years at my institution), s/he will not perform patient care services or practice medicine within the greater metropolitan area. This "non-competition agreement" does not usually apply to faculty who (1) retire, (2) are involuntarily removed due to a negative review, a negative tenure decision, or other circumstances under which the employee is involuntarily removed from employment with approval of the dean; (3) are in fellowship training; or (4) are temporary employees. In the event that the faculty member elects to terminate his or her employment and work in the community, it is customary for that person to buy out the obligation by payment to the medical school of an amount equal to 1 year's current total income plus recruitment, moving and costs incurred by the department to hire and support that faculty member. A waiver or modification of this policy may be granted before hire or once termination is selected if justified by exceptional circumstances. A department chairperson, upon receipt of a request from the faculty member, must submit a written justification to the dean detailing the rationale to support this waiver.

6.7 Costs of Faculty Turnover

Faculty turnover has been part of the management literature for decades [16]. With the retirement of faculty members hired during the 1970s and the impending retirement of baby boomer professors, the issue of turnover has assumed an even greater importance [17]. Much of this literature dealt primarily with psychological, sociological, and economic reasons to explain employee departures.

Only until recently have organizations begun to quantify the direct and indirect costs of employee turnover [18]. In the handful of studies from medical schools, costs of faculty turnover were influenced by the type of department (clinical versus basic science), whether the institution was public or private, and geographic region [19]. Therefore, estimates can vary widely from $113,000 per faculty member in a Southern public medical school to $926,000 in a state-run health care system in the North [20].

Schloss et al. [21] reported at one academic medical center that the financial impact of turnover among faculty in clinical departments was very high, even when lost hospital and research revenues were not included. For example, estimates of the cost of faculty turnover in clinical departments at the University of Arizona were higher than those obtained in other published studies (Table 6.2) [19, 22]. The average annual institutional cost of faculty turnover in the Arizona departments of internal medicine and surgery was more than $400,000.

Table 6.2 Comparison of average faculty replacement cost estimates at selected U.S. colleges of medicine, from a study of faculty turnover at the University of Arizona College of Medicine, FY1999–FY2004

Institution	Clinical faculty
University of Arizona College of Medicine[a]	$400,233
Southern Public Colleges of Medicine[b]	$113,000
Northern Private Colleges of Medicine[b]	$261,000
Southwestern Colleges of Medicine[c]	$154,333–$185,254

[a]*Source*: Schloss et al. [21] with permission; total cost divided by the number of faculty members replaced
[b]*Source*: Wenger [19]
[c]*Source*: Waldman et al. [22]

It is likely that much of the variability in faculty turnover costs between schools may result from the inclusion of different variables in the calculations. Bickel et al. [23] recommended that turnover costs include the following additional expenses: advertising, search firm, and/or administrative costs; interview travel expenses (now becoming more off-campus for leadership positions); faculty and search committee time for interviewing; overtime for other staff during the hiring process; orientation and training time; loss of productivity; and clinical income costs of lost patients, referrals, and grants.

Costs of faculty turnover also vary by rank and specialization. A certain positive aspect of employee turnover can be replacement with more junior faculty with lower salaries and reduced benefits, and with newer perspectives. Despite this, costs of this turnover, however estimated, can be impressive and would warrant substantial efforts to foster the success of faculty worthy to be retained.

6.8 Special Considerations

6.8.1 Gender Equity

Women now comprise about half of medical school graduates, yet most who are faculty are instructors or assistant professors [24]. Any disparities in salary not only affect women currently employed in academic medicine but influence perceptions of their desirability and the next generation to pursue a career in academic medicine. Any salary disparities may also contribute to the higher attrition rates among women faculty [25].

Wright et al. [26] compared adjusted annual salaries for full-time women and men faculty before and after an intervention at their medical school. Gender disparities in compensation can be reduced through careful documentation, identification of comparable individuals paid different salaries, and commitment from leadership to hold the appropriate person accountable. Other measures such as facilitating part-time work, other family-friendly policies, and providing flexibility in tenure clocks are critical to the advancement and retention of current women faculty and the next generation of faculty [27].

6.8.2 Public Versus Private Medical Schools

According to the 2009 Faculty Salary Survey, average salaries of faculty members at public medical schools continue to be lower than for those at private medical schools [8]. An explanation for this finding could be a continued flattening or a decline in state appropriations for public medical schools [10]. Studer-Ellis et al. [7] reported that trends in U.S. medical school faculty salaries between 1988–1989 and 1998–1999 of faculty members at public medical schools tended to grow more slowly than those of faculty members at private medical schools. We are now examining whether average salaries of clinical faculty at community-based medical schools or at community-based residency programs grew at rates higher than salaries for faculty at more traditional schools.

6.8.3 Comparison with Private Practice

The finding reported here that salaries of faculty specializing in women's health care kept up with inflation is encouraging. Between 1995 and 2003, average physician net income from the practice of medicine declined about 7% after adjusting for inflation according to a national study from the Center for Studying Health System Change [28]. Among different types of community practitioners, primary care physicians fared the worst with a 10.2% decline in inflation-adjusted income during this period, while general surgeons' real income declined by 8.2%. Flat or declining fees from both public and private payers appear to be a major factor underlying declining real incomes for physicians. Since the mid-1990s the downward trend in real incomes is likely an important reason for any growing unwillingness by community physicians to assume common responsibilities of faculty such as teaching medical students and undertaking pro bono work, such as charity care and volunteering to serve on hospital committees.

6.9 Conclusions

While actual medical school faculty salaries are increasing, the real growth rate of average clinical faculty salaries is essentially keeping up with inflation. Overall salaries in obstetrics and gynecology faculty are now growing at rates similar to "nonprocedural" women's health care specialties such as family medicine and general internal medicine. Growth in real purchasing power, however, improved very slightly or not at all during this past decade.

Salary contracts require careful and sensitive consideration at the time of recruitment and at annual review. Costs of faculty turnover can be impressive and warrant substantial efforts to foster the success of worthy faculty or to implement a "non-competition" policy when applicable. A growing body of anecdotal evidence suggests that clinical faculty are spending more time in patient care (at the expense of time in teaching and research) to maintain strained departmental budgets. It could be argued, however, that growth in salaries during this period would be much less if faculty members and academic health centers did not make these adjustments. Therefore, performance-based incentive compensation plans that prevail in industry are gaining more attention at academic health centers [29, 30].

During this time of economic unrest and health care reform, it would be prudent to examine the trends described herein more closely and compare salary levels of academic women's health care physicians with their counterparts in private practice. In addition, the influence of gender, tenure, years since appointment, and part-time positions on faculty salaries require periodic consideration. Lastly, future investigations might examine the effects of these salary trends on the retention of clinical faculty, particularly those who are effective teachers or investigators in women's health care.

References

1. Rayburn, W.F., B.L. Anderson, J.V. Johnson, M.A. McReynolds, and J. Schulkin. 2010. Trends in the academic workforce of obstetrics and gynecology. *Obstetrics and Gynecology* 115:141–146.
2. Cromwell, J., W. Adamache, and E. Drozd. 2006. BBA impacts on hospital residents, finances, and Medicare subsidies. *Health Care Financial Review* 28:117–129.
3. Robinson, L., ed. 2009. *AAMC data book: Statistical information related to medical schools and teaching hospitals*. Washington, DC: Association of American Medical Colleges.
4. Smithson, K.W., and J. Koster. 1997. Incentives and the management of physician behavior in health service organizations. *Journal of Ambulatory Care Management* 20:8–16.
5. Feldman, A. 2010. *Pursuing excellence in healthcare: Preserving America's medical centers*. Boca Raton, FL: CRC Press.
6. Krakower, J.Y., D.J. Williams, and R.F. Jones. 1999. Review of US medical school finances, 1997–1998. *JAMA* 282:847–854.
7. Studer-Ellis, E., J.S. Gold, and R.F. Jones. 2000. Trends in US medical school faculty salaries, 1988–1989 to 1998–1999. *JAMA* 284:1130–1135.
8. Alexander, H. 2009. *Report of medical school faculty salaries, 2007–2008*. Washington, DC: Association of American Medical Colleges.
9. Rayburn, W.F., A.M. Fullilove, J.A. Scroggs, and R.M. Schrader. 2011. Trends in salaries of obstetrics-gynecology faculty, 2000–01 to 2008–09. *American Journal of Obstetrics and Gynecology* 204:82.e1–6.
10. 1998. *Implications of the evolving health care system for academic medicine*. Washington, DC: Association of American Medical Colleges.
11. Stites, S., L. Vansaghi, S. Pingleton, G. Cox, and A. Paolo. 2005. Aligning compensation with education: Design and implementation of the educational value unit (EVU) system in an academic internal medicine department. *Academic Medicine* 80:1100–1106.
12. Watson, R.T., and I.J. Romrell. 1999. Mission-based budgeting: Removing a graveyard. *Academic Medicine* 74:627–640.
13. Nutter, D.O., J.S. Bond, B.S. Coller et al. 2000. Measuring faculty effort and contributions in medical education. *Academic Medicine* 75:200–207.

14. Tarquinio, G.T., R.S. Dittus, D.W. Byrne, A. Kaiser, and E.G. Neilson. 2003. Effects of performance-based compensation and faculty track on the clinical activity, research portfolio, and teaching mission of a large academic department of medicine. *Academic Medicine* 78:690–701.
15. Alexander, H., and S. Lang. 2008. Analysis in Brief: The long-term retention and attrition of U.S. medical school faculty. *Association of American Medical Colleges* 8:1–2
16. Cavanaugh, S.J. 1989. Nursing turnover: Literature and methodological critique. *Advanced Nursing* 14:587–596.
17. Bahrami, B. 2001. Factors affecting faculty retirement decisions. *The Social Science Journal* 38:297–305.
18. Hall, T.E. 1981. How to estimate employee turnover costs. *Personnel* 58:43–52.
19. Wenger, D.C.K. 2003. Conducting a cost-benefit analysis of faculty development programs: Its time has come. *Academy of Physical Science* 1:6–7, May/June 2003.
20. Scott, K. 1998. Physician retention plans help reduce costs and optimize revenues. *Health Finance Management* 52:75–78, Jan 1998.
21. Schloss, E.P., D.M. Flanagan, C.L. Culler, and A.L. Wright. 2009. Some hidden costs of faculty turnover in clinical departments in one academic medical center. *Academic Medicine* 84:32–36.
22. Waldman, J.D., F. Kelly, S. Arora, and H.L. Smith. 2004. The shocking cost of turnover in health care. *Health Care Management Review* 29:2–7.
23. Bickel, J., D. Wara, B.F. Atkinson et al. 2002. Increasing women's leadership in academic medicine: Report of the AAMC project implementation committee. *Academic Medicine* 77:1043–1061.
24. Magrane, D., J. Lang, and H. Alexander. 2005. Women in U.S. academic medicine statistics and medical school benchmarking, 2004–05. *Association of American Medical Colleges* 6(7):1–46.
25. Yamagata, H. 2009. Trends in faculty attrition at U.S. medical schools analysis in brief. *Association of American Medical Colleges* 2(2):1–2.
26. Wright, A.L., K. Ryan, P. St Germain, L. Schwindt, R. Sager, and K.L. Reed. 2007. Compensation in academic medicine: Progress toward gender equity. *Journal of General Internal Medicine* 22:1398–1402.
27. Kahn, J., S. Degen, and M. Mansoor. 2005. Pediatric faculty members' attitudes about part-time faculty positions and policies to support part-time faculty: A study at my medical center. *Academic Medicine* 80:931–939.
28. Tu, H.T., and P.B. Ginsburg. 2006. Losing ground: Physician income, 1995–2003. *Tracking Report* 15:1–2.
29. Emery, S.E., and C. Gregory. 2006. Physician incentives for academic productivity: An analysis of orthopaedic department compensation strategies. *Journal of Bone and Joint Surgery* 88:2049–2056.
30. Reece, E.A., O. Nugent, R.P. Wheeler, C.W. Smith, A.J. Hough, and C. Winter. 2008. Adapting industry-style business model to academia in a system of performance-based incentive compensation. *Academic Medicine* 83:76–84.

Chapter 7
Healthcare Reform and its Potential Impact on Academic Women's Healthcare Practice and Training

Eric J. Hodgson and Charles J. Lockwood

Abstract When comprehensive health reform is enacted, the practice of medicine and medical training will change in a more major way than following any failed short-term fixes. This chapter emphasizes changing roles of faculty in scholarly activity and in training the next generation of women's health practitioners.

Keywords Health care reform · Health care delivery system reform · Accountable care organization · Graduate medical education · Obstetrics/gynecology · Medical information technology

7.1 Introduction

The enactment of The Patient Protection and Affordable Care Act (PPACA) on March, 23, 2010, marks the most far-reaching efforts at health care reform in the United States since the passage of the Medicare Act of 1965. Although highly contentious and extremely partisan to the bitter end, this legislation represents the first successful, modern-day attempt to address a financially unsustainable health care system. The primary aim of this bill was to increase access to care by providing health insurance to almost all Americans. A secondary focus was to dampen the escalating costs of health care. Unfortunately, the act does not significantly address cost issues, and the fact remains that the health care provided in the U.S. varies greatly in quality, is inequitably distributed, and is ironically subject to both over- and under-utilization. Ultimately, a complete transformation of the U.S. health care delivery system will be required to make increased access affordable. Because the reforms within the PPACA will be rolled-out over many years and will be subject to the ebb and flow of politics, the final nature of health care delivery reform remains to be determined. Equally uncertain is the precise effect such reform will have on academic medicine in general and on OB/GYN practice and training in particular.

E.J. Hodgson (✉)
Division of Maternal-Fetal Medicine, Department of Obstetrics, Gynecology and Reproductive Sciences, Yale University School of Medicine, New Haven, CT, USA
e-mail: eric.hodgson@yale.edu

W.F. Rayburn, J. Schulkin (eds.), *Changing Landscape of Academic Women's Health Care in the United States*, International Library of Ethics, Law, and the New Medicine 48,
DOI 10.1007/978-94-007-0931-7_7,

The goal of this chapter is to delineate some of the critical problems in our current system that prompted reform, to discuss their effects upon women's health, discuss how the two most likely frameworks for a reformed delivery system might deliver higher quality/less expensive care, and to outline the impact of delivery reform on the practice of OB/GYN and residency training.

7.2 The Excessive Cost of U.S. Health Care

Expenditures for health care in the U.S. continue to expand at an alarming rate. Based on U.S. Senate Finance Committee estimates, our health care system is the most expensive in the world. On average, we spent $6,102 dollars per person in 2008, despite the fact that over 48 million Americans were uninsured at that time [1]. Health care costs, modestly dampened by the recent recession, are still increasing at about 5% per year, outpacing wage growth and inflation. The Center for Medicare and Medicaid Services (CMS) has estimated that U.S. health care spending, currently $2.3 trillion dollars a year, will reach $4 trillion by 2016, accounting for 20% of gross domestic economic output (GDP).

The expanding burden of rising medical care costs falls disproportionately on U.S. industries and employers, inhibiting U.S. economic competitiveness. American businesses cover 50–65% of U.S. health care costs. None of our industrialized competitors burden their industries so significantly. The lack of competitiveness of our auto industry is just one example. General Motors spends more for health care than on steel with an extra $2000 added to the sticker price of every automobile [2]. In 2006, the cost of health care for Starbucks employees matched the cost of its coffee purchases [3].

Defensive medicine, or the provision of medical services to decrease the risk of future malpractice litigation, adds significant cost to U.S. health care. According to a recent study looking at increases in health care expenditures, between $100 billion and $178 billion per year may be attributable to defensive medicine practices [4]. If the additional cost of excessive liability insurance premiums, defense of malpractice cases, insurance and judicial administrative costs and astronomical awards are added to this figure, the amount is staggering. Additionally, the emphasis on avoiding litigation and the practice of defensive medicine clouds the training of OB/GYN residents and may play a role in discouraging medical students from choosing the specialty [5].

7.3 Prior to Health Care Reform Many Americans – Particularly Women – Were Uninsured, Underinsured, or at Risk of Losing Benefits

According to the US Census Bureau in 2007, immediately prior to passage of the PPACA, 85% of Americans had some form of health insurance while at least 15% of Americans were uninsured. Of those with private insurance, nearly 60% obtain it through their employer while 9% purchased individual policies [6]. Various government agencies (e.g., CMS and Veterans Administration) provided coverage for another 28% of Americans.

In economic downturns, those Americans who obtain employee-based benefits are at great risk of losing coverage. According to data from the U.S. Census Bureau and the Bureau of Labor Statistics, it is estimated that 2.4 million workers lost health coverage through their jobs during the recent recession [6, 7]. Prior to passage of the PPACA, women were at much higher risk of being uninsured and more often subject to predatory or discriminatory insurance industry practices. Insurance providers in many states were able to consider gender when setting premium rates in both group and individual insurance markets prior to PPACA. In some states, it was legal for an insurance company to reject a woman's health insurance application for arbitrary and biased preexisting conditions such as having a Cesarean delivery or being the victim of domestic violence. Many women receive insurance through their spouse's employer and were subject to loss of insurance in the event of divorce. Without statutory mandates, maternity care was often not included in individual insurance plans or available only with extremely high deductibles or as a specific rider to the coverage. When a woman was able to purchase individual coverage, she often paid more than a man for identical coverage, often paid higher deductibles, and might have to go without coverage for a wide range of reproductive health options [8]. Many of these inequities were addressed by PPACA such as the inability of an insurance company to deny coverage based on a preexisting condition, inclusion of maternity benefits in the core plans offered in the insurance exchange, the end of rescissions from the individual market where women were more likely to have their coverage cancelled.

Prior to the PPACA, over 50% of all personal bankruptcies cited medical illness or medical expenses as a causative factor [9]. Surprisingly, most of these individuals had health insurance at the start of their illness, although 38% had lost coverage by the time they filed for bankruptcy [9]. Those considered "underinsured," particularly women and those purchasing individual policies, were at greatest risk for bankruptcy and fell into the category of people who are "one serious medical illness away from financial ruin" [9]. Again this abuse has been partially addressed in the reform legislation.

In 2006, about one-fifth of the 120 million emergency department (ED) visits in the U.S. were by the uninsured. Inappropriate ED utilization is expensive and often goes unremunerated [10]. Difficulty in accessing prenatal care prompts many women to utilize the ED for non-emergent, obstetrical issues in order to gain reassurance during an ongoing pregnancy. While PPACA should reduce the number of uninsured using the ED as their primary care provider, it does not address the country's growing lack of primary care providers. Lack of primary care causes delays in seeking treatment until chronic illness has progressed to a more acute stage. Academic health centers (AHCs) currently play a large role in providing care for the uninsured. At the present time, it is estimated that 50% of the uninsured receive medical care through AHCs. AHCs are in a unique position to address the needs of the uninsured given their frequent location within major population centers, commitment to provide care to all, strong social service case management systems that seeks to identify payment mechanisms for uninsured patients, and their ability to absorb hundreds of millions of dollars in charity care [11]. This financial burden should improve after PPACA either because these patients will now be insured or because they will be able to seek care elsewhere.

7.4 Access Does Not Guarantee Quality, Equitable or Safe Care

Even with improved PPACA's likely improvements in *access to care*, the current reform law does not fundamentally address the relatively poor quality of U.S. health care delivery. Although technologically advanced and expensive, the U.S. system underperforms most other developed countries. The USA spends 50–100% more than other industrialized nations on our health care system but ranks 37th according to the World Health Organization (WHO) when considering overall population health, health disparities, health system responsiveness, universal access, and financial burden [12].

Ironically, while many experts cite over-utilization of expensive technology as a primary driver of costs in the U.S. system, there is evidence of substantial underutilization of basic primary preventative and chronic care. A Rand Corporation study indicated that only 50% of Americans receive recommended preventive care; 70% receive recommended acute care and 60% receive recommended chronic care [13]. Ironically the field of OB/GYN is leading the way in improvements in patient safety and quality have been undertaken, in part, to mitigate the financial impact of malpractice litigation. As quality improvement measures have been initiated, the size and complex mission of AHCs make it "orders of magnitude" harder to achieve such improvements than in other healthcare settings [14, 15].

7.5 The US Healthcare System Provides Erratic and Uncoordinated Care

Recent Federal reform legislation also does not fully address geographic, racial and socioeconomic disparities in care including inconsistencies in the type of care provided, the relative cost of that care, and the predominance of medical specialists in a given area [16].

Although the methodology has been questioned [17], oft-cited studies correlating Medicare expenditures with outcome have shown a three-fold difference in spending per patient between regions with the highest and lowest costs without demonstrably better outcomes, higher patient satisfaction, or improved access to care [18, 19]. Thus, providing more Americans with health insurance will not directly improve quality, or patient safety nor control the additional, direct costs incurred from preventable medical errors. For example, inpatient adverse drug administration events (ADEs) increase an affected patient's hospital cost by $4,700 while outpatient ADEs in aggregate account for an extra $77 billion in health care expenditures [18]. Errors also fuel costs indirectly through litigation. Liability insurance premiums cost $6 billion each year and, as noted, fear of litigation and defensive medicine add a minimum of 3–4% to health care expenditures [18, 20].

7.6 Our Health Care Delivery System is Fundamentally Flawed

Current health care reform act focuses on decreasing the number of uninsured but does not address fundamental problems in how care is delivered [18].

1. *Care is uncoordinated*: The U.S. health care system is organized and regulated along traditional academic specialties or silos that often do not effectively communicate. This results in fragmented, uncoordinated care and hinders quality improvement efforts. The lack of a national electronic health record with embedded decision support software is a major cause of duplicated tests, medical errors and inefficiency.
2. *The Fee-for-Service (FFS) payment system creates waste*: Fee-for-service payments lead providers to focus on volume, not health maintenance, efficiency of care or even quality. Conversely, to increase profits, health plans are maximally incentivized to create complex billing and reimbursement rules that delay payments or actually shift costs to patients and providers.
3. *Lack of information about cost and quality perpetuates waste and hampers improvement*: Given the lack of universal reporting of provider outcomes, substandard services can flourish without a means of detection, assessment or quality improvement. Similarly, failure to publicize provider utilization of costly tests, procedures, consultations and other services perpetuates FFS-driven inefficiency.
4. *Focus of U.S. health care is short-term*: Health plans focus on reducing short-term costs given their 12 month selection cycle. This short term focus stifles promotion of preventive care.
5. *Current governmental regulations on health care obstruct innovation*: Regulations such as Stark laws, anti-kickback regulations, etc. sometimes serve to impede innovation within delivery system/funding structures or the introduction of cost-saving or value-enhancing technical innovations.

7.7 Future Reform of the Delivery System: Two Contrasting Models

Given that federal health care reform will increase access into a fundamentally flawed delivery system, truly effective reform will require a major restructuring of the way health care is provided. What, then, are possible delivery prototypes that we may consider and how will delivery system reform change the structure and function of Academic medicine in the next decade?

Two sharply contrasting delivery models have been proposed by Michael Porter from Harvard and Alain Enthoven at Stanford. In their book, *Redefining Health Care*, Porter and his co-author Elizebeth Olmsted Teisberg, propose a radical restructuring of the present health care delivery system where specialized teams or Integrated Practice Units (IPUs), focused on specific medical conditions, compete on the basis of patient outcomes relative to cost [21–24]. The alternative model espoused by Enthoven and associates would create large, integrated health care delivery systems (IHDS) that offer a full range of care and compete on price [25–27].

7.8 Porter and Teisberg and Value-Based Care

The central tenet of the Porter and Teisberg model is that healthy industries continually work to create improved products and services at a lower cost (i.e., value). Those unable to compete on value are driven out of business. Often, leaps in value occur as this positive-sum competition spawns "disruptive" technologies such as the abandonment of mainframe computers following introduction of the personal computer.

In contrast to healthy industries, Porter and Teisberg describes the U.S. health care system as one where entities compete for limited health care dollars and divide up "value" rather than increase it. This zero-sum competition stifles "value added" creativity and generates a complex schema of cost shifting throughout the system. As noted, to limit payments to providers, health plans often create extremely complex billing rules making it almost impossible to be reimbursed for some patient care. In exchange for access to the pool of "in-network" patients, insurance plans often negotiate significantly reduced FFS payment schedules for providers. To recoup this lost revenue, physicians are under pressure to maximize the number of patient encounters and procedures thereby spending less time per patient or softening the indications for visits and procedures. Ultimately, this cost shifting dance results in decreased "health care value" to the patient and an increase in errors, and thus net expense.

In place of the current volume-driven, FFS structure, Porter and Teisberg would create a system that encourages the formation of Integrated Practice Units (IPUs), care delivery entities having a focused expertise in a specific medical condition (e.g., gynecologic cancers or vaginal birth after prior cesarean delivery or menopause). Because outcome and cost information would be universally available, patients would have the ability to select providers based on the published value of their care (i.e., patient health outcomes/per dollar spent over the full cycle of care). This value-based competition (VBC) [22] would promote healthy rivalry between different IPUs as they battle for patients (and therefore reimbursements). Better performers will ultimately be rewarded with more patients and more revenue (see Table 7.1 for VBC principles).

Table 7.1 Value-based care

Competition based on value (better patient outcomes for fewer dollars spent).
Outcomes measured over the full cycle of care.
Value is driven by provider experience and learning at the medical condition level.
Outcome and cost information that supports value-based competition must be widely available.
Innovation that increases value must be strongly rewarded.

Source: Porter and Teisberg [22]

Each specialty IPU would include all requisite components needed to provide focused care including physicians with expertise in a specific medical condition, dedicated facilities and staff. Disease management would require close involvement with patients over an extended period to improve adherence with medication and lifestyle changes, detect impending problems, and implement timely, cost-effective care. Primary care IPUs would focus on disease prevention and early detection (e.g., breast exams, Pap smears) and refer patients to the highest quality medical condition-specific IPU for treatment when needed (e.g., breast surgeon, gyn oncologist, urogynecologist, adolescent gynecologist). Another crucial role for primary care IPUs would be to coordinate overall care and collaborate with specialist IPUs on chronic disease management.

Care reimbursement in this model would be based on the "cycle of care" (COC) for each condition. For example, a COC for a patient with newly diagnosed breast cancer might include the interval from initial diagnosis through the decision to utilize conservative or radical treatment, the actual surgery, and post-treatment chemotherapy. For a chronic condition like type-2 diabetes or severe osteoporosis, the reimbursed COC would be based on discrete intervals of time (e.g., 6 months).

But how does this model control costs? Porter posits that high quality, well-coordinated, specialized care by IPUs in VBC will ultimately be less expensive; providers would develop the requisite experience, scale and efficiencies to make earlier, more accurate diagnoses, would use less invasive and expensive therapies and would better integrate and coordinate care to avoid waste and errors. Newer technology would be embraced as long as it added value (i.e., better outcomes for less cost).

7.9 Enthoven and Competition Based on Integrated Delivery Systems

In contrast, Enthovan and Tollen [26] envision large, regional, fully integrated health care systems (IHDS) as the basic unit of health care delivery. The typical IHDS would include multispecialty medical groups, multiple hospitals, labs, imaging facilities and pharmacies. All services would be reimbursed on a per capita prepayment system (i.e., capitation) or through their own health plan offering (e.g., the Kaiser system). Competition in this model occurs at the system level with consumers choosing an IHDS and not the best individual provider for a given medical condition. Unlike the Porter model's highly individualized, sharply-focused IPU-based VBC model, the Enthoven model is a highly centralized, overarching management system offering a protean array of care. Whereas the Porter model focuses on value, the Enthoven model focuses on cost containment.

Those investigators theorize that only the IHDS -based model can ensure proper oversight of providers to make certain they are competent, well-trained, and proficient in specific evidenced-based diagnoses and treatments. They would devise broad-line quality strategies, similar to General Electric, which would "maintain a rigorous environment of evaluation and discipline based on quality and efficiency." [27]. Conversely, they assert that only IHDS can deploy physicians in appropriate numbers and specialties to adequately meet a population's needs.

Enthoven and associates believe that capitated (per member-per month) global payments reward doctors for health maintenance and prevention, encourage appropriate use of procedures and promote cost-effective problem solving. The IHDS would focus on cost reduction and would adopt only cost-effective technologies thus ending the vicious cycle of escalating procedure volume inherent in the current FFS system. Enthoven and Tollen note that per capita health care costs are 25–30% lower in prepaid group practices compared with FFS practices and that IHDS are more likely to use care teams, employ better IT systems, and implement clinical guidelines [27]. Finally, they contend that outcome reports and quality bonuses will

Table 7.2 Comparison of the current models of health care reform

Model	Porter	Enthoven
Nature and Basis of competition	Value: Outcome in treating a specific medical condition over the full cycle of care (COC) for the lowest price.	Cost: Lowest price to pay per member per month.
Decision maker	Individual patients (consumer) aided by web-based value reports and advice from primary care physicians and health plans	Health plans, large companies (Human Resource officials), officials at government subsidized insurance pools or Medicare and Medicaid
Site of competition	Integrated practice units (IPUs)	Integrated health care delivery systems (IHS)
How choices are made	Consumer seeks highest value IPU for a given medical condition or for primary care based on published outcome and price data aided by health plans and PCPs	Health plan (or large employer or government bureaucrat) based on cost
Coordinator of care	IPU	Primary care provider based on instructions of IHS management
Scope and range of competition	Potentially global	Local and regional

check the propensity to ration care and thus, avoid the criticisms that stifled the widespread embrace of capitation in the late 1990's. For a comparison of the two model's basic principles, see Table 7.2.

7.10 Both Care Delivery Models Require Improved Information Technology (IT) Platforms

Because public reporting of outcomes and cost is fundamental to competition in the Porter model and serves as a check on rationing in the Enthoven model, sophisticated information technology (IT) systems, including integrated electronic health records (EHRs) with comprehensive decision support components, are required to compile requisite information needed to publicly report on patient outcomes, cost, and care methods. These IT systems would merge administrative, clinical, and financial information and could also be linked to patient personalized electronic health records (PHRs) to empower patients as partners in care. These systems will improve documentation, and prevent duplication while enhancing results by allowing integration and coordination of care should reduce errors. In any case, they would guarantee universal provider access to accurate and current patient medical histories, imaging and lab test results.

7.11 The Short-Term Repercussions of Health Care Reform

Regardless of the type of health care delivery system the U.S. evolves towards, PPACA will quickly increase access to care for millions of currently uninsured patients. For women, there will be an elimination of gender-biased premium elevation, removal of the threat of being denied coverage for preexisting conditions, greater ability to afford individual health insurance through tax credits and guaranteed coverage for maternity care. However, overall costs will inevitably increase in the short term as a bolus of newly insured patients creates new demands on an unprepared system. In the immediate future, the Enthoven model seems singularly attractive since it can be more readily implemented, will reduce costs quickly and has the capacity to accommodate larger numbers of patients into existing health systems. Given their large size, IHDS could easily accept patients participating in state or federal government insurance "pools" of individual and small company employees.

From the patient's perspective, fear of care-rationing was foreshadowed by the summer of 2009 "town meetings" held to discuss health care reform options. There overheated rhetoric about "Death Panels" – the fear that group of government bureaucrats would encourage older Americans to forego treatment and die – were used to politicize and polarize the debate. One possible strategy to minimize anxiety over rationing would be to dramatically simplify the process of accessing care by eliminating "gatekeepers" and including well-publicized, pay-for-performance

(P4P) safeguards into all health insurance plans. This would establish a minimal "standard of care" that would make necessary treatments available to all appropriate patients. Moreover, the elimination of an insurance company's ability to deny health insurance for those with a "preexisting condition" should lessen the apprehension for those who currently have health insurance but are afraid that they could someday lose coverage and be unable to qualify for a new plan.

7.12 Long Term Consequences of Reform: A Two-Tiered System

Given the minimal administrative requirements of capitation, and the current provisions that aim to increase the number of those eligible for Medicaid and Medicare, full embrace of the Enthoven model could easily lead to alternative federal or state government payers. However, as was evident during the recent health care reform debate, the intense fear and politicization over the creation of a public option – coupled with the intense lobbying and advertising campaigns of the insurance industry – makes it unlikely that this type of government-sponsored plan will be a viable option in the foreseeable future.

Where does the Porter IPU fit into a newly reformed health care delivery system? Regardless of the assurances of the IHDS leadership, a capitated system will likely provoke fear of care rationing and technology suppression. This fact, coupled with the logistical difficulties patients will likely face when interfacing with the system, will likely spur the development of a "gray" market in health care. In such a market, consumers would pay directly for care from IPU-like providers with expertise in a given specialty or medical condition. Since IPUs are positioned to generate better results than comparable IHDS-based providers and would likely engender far higher patient satisfaction, over time, the entire market could evolve to the Porter model or some mixture of IPUs and IHDS analogous to the U.K.'s National Health Service and its so-called Harley Street private practices.

7.13 Renewing Interest in Primary Care Specialties

Primary care physicians (PCPs) will undoubtedly play a progressively larger role in caring for the newly insured. The current lack of adequate numbers of PCPs poses a serious barrier to the effectiveness of any type of delivery reform. As an example, the State of Massachusetts adopted legislation that provided near universal health care coverage in 2008. Although the State's rate of uninsured is currently at an all time low, there is growing concern that there are not enough PCPs to provide care for this influx of patients [28]. In the broader scheme, it is estimated that adoption of universal health coverage in the U.S. would exacerbate the projected physician shortage of 200,000 in 2020 by 25% [29]. In terms of the OB/GYN workforce, evolving practice patterns within the specialty will likely adapt to the increased demand for ambulatory care provision. As the principal physician for many reproductive age women, OB/GYNs will undoubtedly see an increased demand for their services.

This increased demand may, however, exacerbate the current lack of obstetrics providers in many locations.

In the past decade, the number of U.S. medical school graduates entering family practice residencies and the percent of internal medicine residents planning careers in primary care dropped by more than 50% [30]. According to the American Academy of Family Physicians, U.S. medical school graduates filled only 42% (1,083) of the 2,555 resident positions for family medicine in March 2009. Conversely, medical subspecialty fellowship positions have increased by 40% in the same interval and a quarter of general internists left their practices after only 15–20 years. This trend has also been seen in OB/GYN with the increase in the number of applicants to subspecialty fellowships within our specialty in the recent past. For example, in 1999, the number of subspecialty positions offered within OB/GYN was 114 and efforts were made to encourage OB/GYN residents to pursue fellowship opportunities because of difficulty filling these positions [31]. In contrast, in 2009, there were 328 applicants for 200 available fellowship positions within the four major OB/GYN subspecialties [32].

There are multiple causes for this national PCP collapse. Economic realities make primary care less attractive than the more lucrative subspecialties. Falling incomes coupled with the fact that the average debt for a newly minted physician was $154,607 in 2008 [33] makes choosing a primary care specialty a daunting or even financially untenable option for many. Moreover, there has been a 10% drop in PCP income over the past decade layered onto a far lower base salary than specialists receive ($162,000 versus $297,000) [34]. Under FFS, the quality of work life has deteriorated for PCPs as they attempt to maintain income by seeing ever-greater numbers of patients in the face of ever-falling reimbursements. The average PCP now requires 10.6 h a day to provide high quality long-term care and an additional 7.4 h per day to provide all the recommended preventive services [35]. It should come as no surprise that 42% of PCPs report not having sufficient time for patient care. More importantly, there is no current structure for reimbursement for coordinating care, arguably one of a PCP's most important functions.

Despite these challenges, on average, PCPs provide higher quality care at lower cost. Patients with a PCP as their personal physician compared to those utilizing a specialist have 33% lower health care costs and 19% lower risk adjusted mortality [36]. Furthermore, increased access to PCP is associated with reduced all-cause mortality [37]. Thus, of all the broken aspects of the U.S. health care system, the devastation of primary care is perhaps the most serious.

7.14 Patient Centered Medical Home

One possible solution to address the primary care crisis is implementation of the patient centered medical home (PCMH). The core principles of the PCMH are based upon providing personal, patient-oriented, coordinated care through an integrated

Table 7.3 Patient centered medical home requirements

(1)	Provision of primary care including coordination of care, preventative care, health maintenance and acute health care services.
(2)	Electronic health record with decision support capabilities. This system would be leveraged to generate timely reports listing covered patients and resources used for P4P payments or penalties.
(3)	Quality assurance and improvement program.
(4)	24 h access and/or communication.
(5)	Up to date advanced directives.
(6)	Contract with each patient designating provider as their medical home.
(7)	Payment by capitation or combination of monthly payment for medical home infrastructure (e.g., IT) and care coordination plus FFS for discrete care provided modified by P4P criteria.
(8)	Care coordinators to assist patient in self-management, enacting life-style changes and monitoring progress
(9)	Medication monitoring (necessity, dosage, potential for adverse drug reactions)

Source: Starfield et al. [37]

team of professionals. Quality and safety are paramount in this construct with ultimate responsibility for the health of a patient resting in the hands of the providers of care as opposed to insurance companies. A hallmark of the PCMH is the empowerment of the patient to achieve health through increased access to the care team. The PCMH is designed to maximize preventive care, wellness programs and care coordination by fairly compensating these previously non-reimbursed services. This could address two divergent problems: improved reimbursement of PCPs while simultaneously improving care provision through better coordination and oversight. PCMH will likely reduce costs by decreasing ED visits, prevent duplicated imaging, hospital readmissions and unnecessary procedures. The potential savings and improved safety outcomes are substantial [38]. While the nature of the exact structure of a PCMH is under debate, minimal requirements have been outlined by the Medicare Payment Advisory Committee (MedPAC) (see Table 7.3) [39].

7.15 Accountable Care Organizations (ACO)

Similar in structure and intent to the PCMH is an Accountable Care Organization (ACO). An ACO is a local health care organization with a team of providers (at a minimum, primary care physicians, specialists, and hospitals) that would be held accountable for both the cost and quality of care delivered to a defined population. The primary goal of the ACO is to deliver coordinated and efficient care that would achieve specified quality and cost targets. Successful ACOs would be rewarded with a financial bonus; under some approaches, those that fail might be subject to a financial penalty. In order to meet the requirements of this type of incentive system, an ACO needs to be able to:

- Care for patients across the continuum of care, in different settings.
- Plan, prospectively, for its budgets and resource needs.
- Support comprehensive, valid and reliable measurement of its performance.

The ACO ensures that the people and organizations that actually provide care are accountable for the quality and the cost of that care. Previous health reform initiatives involved insurers and made them ultimately accountable. The concept driving ACOs is that it is providers, not insurers, who are best placed to make the changes that will address the cost and quality problems resulting from the U.S.'s current system of fragmented care, variation in practice patterns and volume-based payment systems [40].

7.16 Retail Clinics

Another strategy to enhance primary care post-reform is to decant some of the volume from a PCPs primary office and EDs to facilities that provide care using simple, rule-based diagnostic and treatment algorithms (e.g., retail-based clinics). Simple problems, such as urinary tract infections, strep throat, pregnancy tests, and even routine vaccinations can easily be managed in a "walk-in setting" located at a shopping mall or retail pharmacy. Such an urgent care setting could easily be incorporated into a PCMH and therefore continue to provide coordinated services.

7.17 The First Steps to Reform are Already Underway

Prior to PPACA, steps had already been undertaken to address some of these concerns and to foster an environment where either the Enthoven or Porter model of care could be realized. Demonstration projects are being funded by the CMS P4P project and through its Quality Monitoring System. MedPAC is now proposing large scale testing of PCMHs paid for with a mix of per capita and FFS reimbursement [39]. The former would cover the cost of coordinating care and IT investments while the latter would cover unique services rendered. Current reform efforts contain funding for pilot programs that would move away from the FFS model and to create innovative payment schemes that reward quality and value-based care.

MedPAC is also proposing bundled physician-hospital payments for episodes of care [39]. For example, there would be a single payment for the care of a patient with CHF covering all hospital and provider services both for inpatient care and post-discharge ambulatory care for 60 days. Given that private health plans usually follow CMS's lead, PCMHs and bundled payments will likely be the norm throughout the health care system within 3–5 years. Moreover, these MedPAC proposals could easily serve as a starting point for the creation and integration for either of the discussed models.

7.18 Medical Workforce Reform: Encouraging Medical Students to Pursue Needed Specialties

To encourage medical students to pursue a career in general primary care and related specialties such as ambulatory Ob/Gyn, steps must be taken to diminish the disparity between salaries of specialist vs. generalist physicians. Such measures have been included within PPACA and hope to begin increasing reimbursement of primary care office visits by 5–10% [28].

To offset the burgeoning debt load of medical students, loan repayment programs, such as the National Health Service Corps, could be expanded, be made more flexible and easily available for those willing to practice general primary care, general surgical specialties, and/or in underserved areas. It is also possible to create a medical education system similar to military physicians where medical education is free but requires a mandatory period of service.

As the AAMC works to accomplish the 30% increase in the number of medical schools and medical students by 2015, innovative strategies to attract those students who might be more likely to enter primary care could be used. For example, medical students coming from more rural areas are more likely to practice primary care and should be encouraged and tooled to succeed [41]. As new medical schools are established, it is likely that many will have a specific focus on primary care training and research [42].

7.19 Changes in Graduate Medical Education (GME) and Continuing Medical Education (CME) are Necessary to Prepare Doctors for a Reformed Health Care System

Specific changes in how all physicians are trained will be necessary to meet the needs of a reformed health care system. Although implementation of advanced IT will be critical to improve coordination of patient care and to decrease errors, specific preparation will be needed to utilize these new tools. To practice evidence-based medicine, physicians will require increased training in medical informatics as it pertains to both inpatient and office-based care. To remain current, CME opportunities must be flexible, readily available at low-cost and should optimize use of ever-evolving cyber-technologies.

Despite the fact that that majority of health care is provided in the office setting, the majority of all GME training occurs in highly specialized, inpatient facilities. A primary reason for this imbalance is that current mechanisms for GME do not allow reimbursement for training in community-based sites. Training opportunities that balance inpatient care with training in outpatient, rural, and community sites should be adequately funded and outfitted. "Teaching Health Centers" could focus upon training residents the skills needed to adequately function within a team-based and interdisciplinary care structure [28].

7.20 Reform in the Funding of GME

GME is funded via two primary mechanisms. The largest revenue source, approximately $12 billion per year, is through CMS. This funding source is paid in a few ways. Direct Medical Education Payments (DME) cover salary, fringe benefits, attending physician salaries and is calculated based on the number of Medicare patients and the number of resident physicians within a teaching location. Indirect Medical Education payments (IME) is calculated as a "percent add on" to the DME. These complex formulas are employed to pay higher reimbursement rates to teaching hospitals in order to offset the higher costs associated with training residents. The second major source of funding is through the Health Resources and Services Administration (HRSA) programs that allocate about $530 million at the discretion of Congress. Funding by HRSA supports primary care, nursing, dentistry, and makes available grants and incentives for providers to work in medically underserved communities and in understaffed specialties such as primary care [43].

Unfortunately, these funding mechanisms have not been revised in decades resulting in little relationship between funding of teaching hospitals and the actual cost of GME. Additionally, the number of Medicare-funded GME positions designated for a training site is static after being capped in 1997 as part of the Balanced Budget Act of 1997. Hospitals that have closed specific programs (i.e., pediatrics residency programs) still maintain the same number of paid "slots" regardless if they are being used. Redistribution of these unused spots along with increasing the number of training positions is a critical step in addressing the workforce issues and has been addressed as a provision in the PPACA. A rational approach to GME funding reform may stimulate the creation of innovative training opportunities and facilities, such as Teaching Health Centers, and could help to redistribute the existing physician workforce and better match the locations of training facilities to underserved areas [43].

7.21 Long-Term Health Care Reform and the Practice of OB/GYN

Fortunately, ob/gyns are in a uniquely advantageous position to develop the opportunities and overcome the obstacles created by a reformed health system. In obstetrics, payments already cover a complete COC. Ob/gyn practices are also positioned to serve as a women's care centered medical home, either alone or potentially partnered with internists in a limited multispecialty group practice and would be an indispensible component of an ACO. Indeed, if health care moves to the Enthoven model, ob/gyn practices will need to become affiliated with or owned by large IHDS. In this model, "laborists" would likely dominate inpatient care. Alternatively, there may be increased integration of mid-level providers into large ob/gyn practices where obstetric care will be provided by midwives with physician oversight. In this model, it is likely that the physicians will be salaried employees of a large corporation, similar to currently existing HMO's. For a group that chose to remain independent from an IHDS, the ob/gyn group might be subcontracted by the IHDS. Payments may accrue by straight capitation on a per patient basis or possibly through a modified FFS model. Within an IHDS, ob/gyn subspecialists could operate as similar sub-groups or as employees of the overall IHDS.

Alternatively, if the Porter model dominates, single specialty/sub-specialty ob/gyn groups already represent nascent IPUs. Their full evolution to IPUs would require developing expertise in a distinct clinical focus and achieving outstanding outcomes for a competitive price. A focus on results measurement, reporting capabilities and other IT functions as well as access to needed ancillary and consultative services will be paramount to success.

As a specialty, ob/gyn has been on the forefront of working towards tort reform. Unfortunately, there is little likelihood of such legal reform in the near future. Evolution towards high-quality IPUs would likely reduce errors and improve patient-provider rapport, a key to avoiding litigation. In the advent of evolution to the Enthoven model, it is possible that the government – the largest purchaser and provider of health care – would be forced towards tort reform to reduce tax loads.

7.22 Health Care Reform and OB/GYN GME

Regardless of the type of reform that is ultimately enacted, the training of future ob/gyns will need to be adapted. Initially, with the influx of formerly uninsured patients one would anticipate an increase in the volume and variety of surgical and ambulatory care experiences, particularly in the outpatient setting. Although this may be initially welcomed, it may ultimately hinder the training of future ob/gyns as the balance is tipped towards service and away from education as care is provided for this flood of patients.

Although the number of GME positions is projected to increase, it is unclear how this will specifically impact ob/gyn education, a system already under considerable strain as it attempts to both train excellent clinicians and stay within the confines of stringent duty hour restrictions. Concurrent with the impetus towards primary care and the creation of the medical home, it is possible that this strain upon ob/gyn GME will eventually give rise to the subspecialty of Ob/Gyn Ambulatory Women's Health and the novel training opportunities this may spawn as ob/gyns are integrated into the interdisciplinary teams found in Teaching Health Centers. Obviously, these changes would be dependent upon a rational and overarching reform of the GME funding system.

As the delivery system evolves, we would anticipate modification in the structure and function of ob/gyn residency programs. In either case there will be a focus on prevention and early diagnosis. In an Enthoven-like model of care, it is likely that some semblance of the traditional system of residency and fellowship training might be maintained since a large system may value and focus upon training generalist ob/gyns to provide care within an IHDS. Within a Porter-like system, where specializing would be the norm, early "tracking" into subspecialty training, or focusing specifically on obstetrics or gynecologic training, may become the norm. Likely, a highly specialized group of research-oriented trainees in ob/gyn would "track" into an academic-research oriented pathway and the research focus of traditional subspecialty fellowships would diminish. Regardless, it is unlikely that the current standardized and uniformly administered system of training, where every ob/gyn resident is expected to have a relatively identical training experience, would be maintained.

7.23 Health Care Reform and Women's Reproductive Health

Politics has always played a large role in the practice of ob/gyn, particularly in the area of women's reproductive health. PPAVA prevents any insurance plan that receives money from the federal government to pay for pregnancy termination. Given the current difficulty in receiving abortion training, particularly for those procedures within the second trimester, it would be anticipated that this would worsen an already serious situation [44].

7.24 Conclusions

Churchill said after the British victory at El Alamein "this is not the end. It is not even the beginning of the end. But it is, perhaps, the end of the beginning." Similarly, the PPACA marks the beginning not the end of the long delayed evolution of the U.S. health system. As each aspect of comprehensive health care is evaluated and refined, the practice of medicine and the ways we train our medical workforce will change in a far more radical way than following any of the failed short-term fixes generated over the past 35 years. There will be a significant decrease in the number of uninsured Americans with a concurrent stress upon our current system to care for this large bolus of patients. The system will undoubtedly evolve and change, likely along the two models of care proposed by either Porter and Teisberg or Enthoven and colleagues or both. As our healthcare system matures, a greater emphasis will be placed upon training and retaining PCPs, and novel ways to diminish the burden of medical educational debt must be created. In ob/gyn, GME will need to become more adaptable and flexible to meet the needs of the changing population of patients. Health care reform offers huge opportunities to improve health care for our patients and our communities and to improve the physician's work life. Like all change, the transition will not be easy and is likely to be disruptive in unexpected ways.

References

1. Kazzi, N., More Americans are losing health insurance every day: An analysis of health coverage losses during the recession. *Center for American progress*, 2009. http://www.americanprogress.org/issues/2009/05/insurance_loss.html.
2. Johnson, T., Healthcare costs and U.S. competitiveness. *Council on Foreign Relations,* January 2010. http://www.cfr.org/publication/13325/healthcare_costs_and_us_competitiveness.html
3. Lipsky, M., Outside view: Starbucks and healthcare. *UPI Health Business,* 2006.
4. Kessler, D.P., and M.B. McClellan. Do doctors practice defensive medicine? *The Quarterly Review of Economics and Finance*. http://ssrn.com/abstract=5060
5. Gariti, D.L., T.W. Zollinger, and K.Y. Look. 2005. Factors detracting students from applying for an obstetrics and gynecology residency. *American Journal of Obstetrics & Gynecology* 193(1):289–293, Jul 2005.
6. DeNavas-Walt, C., B. Proctor, and J. Smith. 2009. *Income, poverty, and health insurance coverage in the United States:2008*. Washington, DC: U.S. Census Bureau.
7. Interactive map: Dramatic increase in the uninsured rate in every state. 2009. Center for American Progress, 4 May 2009. http://www.americanprogress.org/issues/2009/05/working_uninsured_map.html
8. Women need health reform. 2010. *Reform matters-making real progress for women and healthcare*. Washington, DC: National Women's Law Center, Mar 2010. http://nwlc.org/reformmatters/pdf/statewhywomenneedhcr/NATIONALHCRFactSheet_FinalPush.pdf
9. Himmelstein, D.U., E. Warren, D. Thorne, and S. Woolhandler. 2005. Illness and injury as contributors to bankruptcy. *Health Affairs* 2005 Jan–Jun(Suppl Web Exclusives):W5,63–W5,73.
10. Abelson, R., Uninsured put a strain on hospitals, *The New York Times*, December 8, 2008. http://www.nytimes.com/2008/12/09/business/09emergency.html
11. Kelley, W. 2009. Academic medicine and real healthcare reform. *The Journal of Clinical Investigation* 119(10):2852–2852. doi:10.1172/JCI4098611.
12. Schuster, M.A., E.A. McGlynn, and R.H. Brook. 2005. How good is the quality of health care in the United States? 1998. *Milbank Quarterly* 83(4):843–895.
13. McGlynn, E.A., S.M. Asch, J. Adams, J. Keesey, J. Hicks, A. DeCristofaro, and E.A. Kerr. 2003. The quality of health care delivered to adults in the United States. *New England Journal of Medicine* 348:2635–2645.
14. Pettker, C.M., S.F. Thung, E.R. Norwitz, C.S. Buhimschi, C.A. Raab, J.A. Copel et al. 2009. Impact of a comprehensive patient safety strategy on obstetric adverse events. *American Journal of Obstetrics & Gynecology* 200(5):492.e1,492.e8, May 2009.
15. Removing Obstacles to Quality Improvement for Academic Health Centers. 2004. *Robert Wood Johnson Foundation,* July 2004. http://www.rwjf.org/reports/grr/047103.htm
16. Gawande, A. 2009. The cost conundrum: What a Texas town can teach us about health care, *The New Yorker Magazine*, 1 June 2009. http://www.newyorker.com/reporting/2009/06/01/090601fa_fact_gawande
17. Bach, P.B. 2010. A map to bad policy – hospital efficiency measures in the Dartmouth atlas. *New England Journal of Medicine* 362(7):569–573 (2010 discussion p 574; Feb 18).
18. Porter, M.E., and E.O. Teisberg. 2006. *Redefining health care*. Boston, MA: Harvard Business School Press, Chapter 1.
19. Fowler Jr, F.J., P.M. Gallagher, D.L. Anthony, K. Larsen, and J.S. Skinner. 2008. Relationship between regional per capita medicare expenditures and patient perceptions of quality of care. *JAMA* 299(20):2406–2412, May 28 2008.
20. Hellinger, F.J., and W.E. Encinosa. 2006. The impact of state laws limiting malpractice damage awards on health care expenditures. *American Journal of Public Health* 6(8):1375–1381, Aug 2006.

21. Porter, M.E., and E.O. Teisberg. 2006. *Redefining Health Care*. Boston, MA: Harvard Business School Press, Chapter 2.
22. Porter, M.E., and E.O. Teisberg. 2006. *Redefining health care*. Boston, MA: Harvard Business School Press, Chapter 4.
23. Porter, M.E., and E.O. Teisberg. 2006. *Redefining health care*. Boston, MA: Harvard Business School Press, Chapter 5.
24. Porter, M.E., and E.O. Teisberg. 2006. *Redefining health care*. Boston, MA: Harvard Business School Press, Chapter 6.
25. Enthoven, A.C., and W.P. Van de Ven. 2007. Going Dutch – managed-competition health insurance in the Netherlands. *New England Journal of Medicine* 357(24):2421–2423, 13 Dec 2007. http://ovidsp.ovid.com/ovidweb.cgi?T=JS&NEWS=N&PAGE=fulltext&AN=18077805&D=medl.
26. Enthoven, A.C., and L.A. Tollen. 2005. Competition in health care: It takes systems to pursue quality and efficiency. *Health Affairs* 2005 Jul–Dec(Suppl Web Exclusives):W5,420–W5,433. http://ovidsp.ovid.com/ovidweb.cgi?T=JS&NEWS=N&PAGE=fulltext&AN=16148024&D=medl.
27. Enthoven, A.C., F.J. Crosson, and S.M. Shortell. 2007. 'Redefining health care': Medical homes or archipelagos to navigate? *Health Affairs* 26(5):1366–1372, Sep–Oct 2007.
28. Whelan, E.M. 2004. Expanding the primary care workforce: An essential element of health care reform, *Center for American Progress,* July 2004. http://www.americanprogress.org/issues/2009/11/primary_care_shortage.html
29. The complexities of physician supply and demand: Projections through 2025. Association of American Medical Colleges. http://www.aamc.org/workforce/.
30. Sepulveda, M.J., T. Bodenheimer, and P. Grundy. 2008. Primary care: Can it solve employers' health care dilemma? *Health Affairs* 27(1):151–158, Jan–Feb 2008.
31. Pearse, W.H., N.F. Gant, and A.P. Hagner. 2000. Workforce projections for subspecialists in obstetrics and gynecology. *Obstetrics and Gynecology* 95(2):312–314, Feb 2000.
32. Results and Data, Specialties Matching Service, 2009 Appointment Year, National Residency Matching Program, Washington, DC, 2009. http://www.nrmp.org/data/resultsanddatasms2009.pdf
33. Steinbrook, R. 2008. Medical student debt – is there a limit? *New England Journal of Medicine* 359(25):2629–2632, 18 Dec 2008.
34. Rosenblatt, R.A., C.H. Andrilla, T. Curtin, and L.G. Hart. 2006. Shortages of medical personnel at community health centers: Implications for planned expansion. *JAMA* 295(9):1042–1049, 1 Mar 2006.
35. Bodenheimer, T. 2008. Coordinating care – a perilous journey through the health care system. *New England Journal of Medicine* 358(10):1064–1071, 6 Mar 2008.
36. Franks, P., and K. Fiscella.1998. Primary care physicians and specialists as personal physicians- health care expenditures and mortality experience. *Journal of Family Practice* 47(2):105–109, Aug 1998.
37. Starfield, B., L. Shi, and J. Macinko. 2005. Contribution of primary care to health systems and health. *Milbank Quarterly* 83(3):457–502.
38. Erickson, S., G. Mansueto, J. Schilz, J. Swanson, and S. Bleeks. 2009. *Proof in practice: A compilation of patient centered medical home pilot and demonstration projects*. Washington, DC: Patient Centered Primary Care Collaborative. http://www.pcpcc.net/pilot-guide
39. Report to the congress: Reforming the delivery system. 2008. MedPAC, June 2008.
40. Devers, K., and R. Berenson. 2009. *Can accountable care organizations improve the value of health care by solving the cost and quality quandaries?* Princeton, NJ: Robert Wood Johnson Foundation, Oct 2009. http://www.rwjf.org/pr/product.jsp?id=50609
41. Phillips, R.L., M.S. Dodoo, S. Petterson, I. Xierali, A. Bazemore, B. Teevan, K. Bennett, C. Legagneur, J. Rudd, and J. Phillips. 2009. *Specialty and geographic distribution of the physician workforce: What influences medical student and resident choices*? Washington, DC: The Robert Graham Center: Policy Studies in Family Medicine and Primary Care.

42. Legislative proposal for creative primary care workforce solutions 2009. American Medical Student Association, 2010. http://www.amsa.org/AMSA/Homepage/About/Committees/StudentLife/PCSolutions.aspx
43. Derksen, D., and E.M. Whelan. 2010. Closing the health care workforce gap: Reforming federal health care workforce policies to meet the needs of the 21st century. Washington, DC: Center for American Progress, Jan 2010. http://www.americanprogress.org/issues/2010/01/pdf/health_care_workforce_execsumm.pdf
44. Annas, G.J. 2009. Abortion politics and health insurance reform. *New England Journal of Medicine* 361(27):2589–2591, 31 Dec 2009.

Chapter 8
Faculty Satisfaction and Retention in Obstetrics and Gynecology

Colleen M. Horan and Louis Weinstein

Abstract Given the resources that medical schools expend in developing their faculty, the academic medicine community has a longstanding interest in faculty satisfaction and retention. This chapter investigates retention rates of first-time assistant professors in obstetrics and gynecology (i.e., faculty with no prior appointments at or above the rank of assistant professor) compared with other core clinical specialties.

Keywords Faculty satisfaction · Faculty retention

8.1 Background

Over the past decade, there has been a significant amount of attention given to the issue of physician supply. The discussion in the literature has pondered whether there exists a general physician shortage, a shortage of primary care or specialist physicians, or an uneven geographic distribution of available physicians [1–5]. Debate has centered largely over whether to increase the supply of physicians by increasing medical school enrollment (class of 2,013 at an all time high of 18,390 students) by increasing the number of graduate medical education programs (which has been stable for 10 years), or by selectively recruiting into certain medical disciplines [6, 7]. This discussion has centered around the issue of physician supply, or improving the production of physicians, with less attention paid to the subjects of attending physician work hours, physician retention and faculty attrition [4, 8]. A different suggestion has been to better use resources by improving retention of the current physician workforce before investing in the training of new physicians [2–5, 8]. The potential cost-effectiveness of such an approach is highlighted by the estimated cost of $300,000 to replace a burned out physician and the calculation that the cost of faculty attrition can account for 5% of the annual budget of an academic medical center (AMC) [9, 10].

L. Weinstein (✉)
Department of Obstetrics and Gynecology, Jefferson Medical College, Thomas Jefferson University, Philadelphia, PA 19107, USA
e-mail: Louis.weinstein@jefferson.edu

W.F. Rayburn, J. Schulkin (eds.), *Changing Landscape of Academic Women's Health Care in the United States*, International Library of Ethics, Law, and the New Medicine 48,
DOI 10.1007/978-94-007-0931-7_8,

The medical literature has been awash with proclamations of an epidemic of physician dissatisfaction. Certain fields seem to be particularly at risk, including obstetrics and gynecology and academic medicine. Dissatisfaction among academic obstetrician/gynecologists may have profound effects on the retention of faculty as well as the recruitment of medical students into this field. Medical school faculty serve as important role models for medical students and can influence their choice of specialty. The issue of faculty satisfaction is paramount to the goal of recruiting the best and brightest medical students into obstetrics and gynecology as well as to the goal of retaining the commitment of those who are currently training the next generation of physicians and performing research that will affect future practice. In 2007, 72% of obstetric/gynecologic residency positions were filled by U.S medical school graduates, compared to 86.3% in 1997 [11]. At the same time, data is showing that a controllable lifestyle is becoming an increasingly important factor in the specialty choices of medical students [12].

The field of medicine as a whole is being affected by changes in practice patterns as well as by changes in physician demographics. These changes are particularly relevant to the practice of obstetrics and gynecology in an academic medical center, where the majority of faculty are junior in rank. According to the American Association of Medical Colleges (AAMC), the percentage of MD entry-level assistant professors leaving academic medicine within a 10 year time period is 44%, which is much higher than the 38% of all faculty who leave [13]. Autry and colleagues published a study of current and previous faculty in obstetrics and gynecology to explore reasons for increased faculty attrition. They found no difference associated with gender, partner status, number of children or ethnicity and the intention to leave one's job in the next 2 years. However, they found that junior faculty and generalists were more likely to have left academics than senior faculty and subspecialists. Physicians who expressed a 50% or higher likelihood of leaving their job within 2 years were less satisfied and were more likely to express frustration with balancing personal and professional life. The most common reason for leaving was the desire to spend more time with family and friends [14].

This chapter will address the topic of satisfaction and retention of academic faculty in obstetrics and gynecology. It begins with a review of factors that affect the satisfaction of physicians in general, satisfaction of the practicing obstetrician/gynecologist, and satisfaction of academic medical center (AMC) faculty. The implications of two important shifts in physician demographics (generational and gender) will be examined. Finally, strategies for improving the satisfaction, retention and recruitment of physicians from this new demographic to the practice of an academic obstetrician/gynecologist will be explored.

8.2 Physician Satisfaction in General

Several sources have identified increasing dissatisfaction among physicians over the past 50 years. This has been attributed to factors such as managed care, lack of time to accomplish required tasks, expansion of non-medical duties, loss of autonomy and prestige, unmet expectations, and professional liability issues [15, 16]. Physicians report a lack of time to accomplish all the necessary administrative, regulatory, and financial tasks required while trying to meet the high clinical expectations of both themselves and their patients [15]. Surveys indicate that increasing numbers of physicians are less than fully satisfied with the practice of medicine, would not choose medicine as a career if they could start over and would not recommend the medical profession to their children or to qualified college students [15]. A 2001 survey of physicians found that 87% reported a decline in physician morale [15]. Another survey found that 17% of physicians rate isolation from colleagues as a moderate or serious problem and that 26.6% are somewhat (21.2%) or very (5.4%) dissatisfied with the practice of medicine [17]. According to one review, dissatisfaction results from the sharp discrepancies between what patients demand and what doctors can accomplish, between the standards set in training and the compromises forced in practice, between the practice of 50 years ago and that of today, and between the "wide open promise of medical science and the limited results of individual doctors" [15].

Solomon and colleagues reported a decrease in physician satisfaction associated with an increase in influence of managed care due to limits on time spent with patients and requirements by insurers, hospitals, and practice directors to see a higher volume of patients in less time [18]. However, since physicians find time spent with patients to have inherent value they find ways to get around these time constraints. Many physicians willingly disregard time limits when they deem it necessary for patient care because this increases their overall satisfaction, even while acknowledging that this decision decreases their compensation [18]. A different solution has been to establish concierge practices, limiting the role of managed care insurers but also limiting the practice to patients who are able to pay for this service [18]. A survey of physicians in Switzerland addressed five dimensions of satisfaction including patient care, work-related burden, income-prestige, personal rewards, and professional relationships with colleagues. Physicians were found to be more satisfied with patient care, professional relationships and personal rewards and least satisfied with work-related burden and income-prestige. Time spent on administrative tasks was found to have a globally negative effect while time spent on continuing medical education had a globally positive effect [19].

In a cross-sectional survey of a random sample of U.S. physicians stratified by age, specialty, and gender, personal accomplishment and emotional resistance were found to be strongly associated with career satisfaction but work-life balance was not found to be a predictor of career satisfaction. The strongest predictor of work-life balance was having control over schedule and hours worked followed by total weekly hours worked [20]. The number of children at home had the same impact on both male and female physicians, with a significant inverse relationship between the number of children at home and work-life balance and emotional exhaustion [20].

It has been questioned whether physician dissatisfaction leads to withdrawal from clinical practice. In one study, the self-reported intent to leave clinical practice was only 35.4% predictive of actual departure from practice 3 years later. Age 55–64 and dissatisfaction were associated with intent to leave clinical practice, but only age was associated with actual departure from practice [8]. However, another prospective study found that physicians who report being strongly dissatisfied were significantly more likely to retire (OR 2.34, 95% CI 1.6–3.5) or to reduce hours (OR 3.6, 95% CI 2.32–5.6) [21].

8.3 Satisfaction for the Obstetrician Gynecologist

In a cross-sectional study of physician career satisfaction involving 42 specialties, obstetrics and gynecology ranked as the third least satisfied specialty [22]. In this study, satisfaction was significantly and positively associated with income and employment in a medical school but negatively associated with working over 50 h per week, being full-owner of a practice, and having greater reliance on managed care revenue [22]. This study also found that control over work hours was associated with more work-life balance, emotional resilience, and personal accomplishment. Specialties associated with "uncontrollable" lifestyle, such as obstetrics and gynecology, were negatively associated with satisfaction [22]. Nonetheless, a similar cross-sectional study found that practicing obstetrician/gynecologists have higher levels of personal accomplishment and work life balance than general surgeons, and more career satisfaction than general internists [20]. Survey data indicates that surgery and deliveries are among the most satisfying aspects of the practice of obstetrics/gynecology while on-call and in-office time are the least satisfying. There is also a significant inverse correlation between overall career satisfaction and total number of hours worked ($p < 0.001$) [23].

A 1987 report from the American Medical Association (AMA) council on long range planning and development predicted that two factors with the potential to affect the organization of an obstetric/gynecologic practice were the high numbers of women entering the specialty and increasing practice expenses (largely associated with liability risks) [24]. Over 20 years later, this prediction has proven accurate. The numbers of women entering the field of obstetrics and gynecology continues to increase. Female physicians on average work fewer hours and retire 5.5 years earlier than male physicians [20, 21]. Professional liability concerns continue to plague the specialty. The 1987 report predicted that diagnostic technology may actually increase litigation as patients may come to expect perfect outcomes in all cases. Twenty years later, Bryan Cowan's presidential address to the Central Association of Obstetricians and Gynecologists stated that "While the rates of maternal and neonatal complications are low, patients have developed an unrealistic expectation of "zero" complications during a pregnancy, expecting nothing less than a perfect infant" [10].

A 2001 survey of practicing American College of Obstetricians and Gynecologists (ACOG) fellows found that the career pressure with which clinicians were most concerned was the extent to which the cost of liability insurance "will affect the duration of my career in obstetrics and gynecology". Clinicians who were most pressured by this issue were more likely to work in solo practice, university settings, or partnerships and less likely to work in health maintenance organizations or the military [23]. Obstetrican/gynecologists who feel that fear of litigation affects their practice have been found to be less satisfied, and 60% of those over age 50 who retired earlier than expected did so due to professional liability costs [11].

It seems impossible to underestimate the impact of professional liability on obstetrician/gynecologists. According to the 2009 ACOG survey on professional liability, 90.5% of respondents had at least one claim filed against them during their careers, with an average of 2.69 claims per physician. In addition, 59.2% of respondents reported making practice changes due to the affordability or availability of professional liability insurance. Approximately one in five respondents reported increasing the number of cesarean deliveries, ceasing to offer a trial of labor to patients who have had a previous cesarean section, and decreasing their numbers of high-risk obstetric patients. One in ten decreased the total number of deliveries performed and 6.5% stopped practicing obstetrics all together. Over 10% of respondents reported that they had liquidated holdings, accessed savings, or secured a loan to fund liability insurance premium increases [25]. These numbers do not include the estimated costs incurred by defensive medicine practices.

New chairs of academic obstetrics and gynecology departments report financial issues as major stressors, identifying human resources, finances, and relationships with school leaders as areas of greatest need for mentoring. Nonetheless, they report high levels of personal accomplishment and satisfaction with their decision to become chairs [26]. The financial stressors felt by chairs can have a trickle down effect on the satisfaction of clinical faculty in the AMC, who are finding that they are valued in terms of billable events rather than for achieving excellence in teaching and clinical care [27].

8.4 Faculty Satisfaction

The greatest resource of an AMC is its faculty, and faculty turnover comes at a huge cost to a medical school. The cost of replacing a clinical faculty member at an AMC has been estimated to range from $115,554 for a generalist to $587,125 for a surgical subspecialist [28]. These costs include lost clinical income, recruitment and hiring expenses, but not loss of revenue to the hospital or the costs of recruiting the significant other of the new hires.

The factors associated with job satisfaction for the obstetrician/gynecologist in academic medicine have been shown to be different from those in private practice. The job aspects that most contribute to satisfaction for private physicians compared to academic physicians are autonomy, physician-patient relationship and insurance reimbursement ($p < 0.01$). For academic physicians compared to private physicians, the aspects most contributing to job satisfaction are teaching, research, and practice variety ($p < 0.01$) [29]. However, academic physicians are coming under increasing pressure to support their departments by clinical revenue. With less grant money available for research, the role of the clinical educator is expanding at AMC. The need to generate more clinical revenue leads to less time and money for teaching and research, the aspects of practice that academic obstetrician/gynecologists find most rewarding. Both male and female academic surgeons report feeling that they are not meeting academic productivity expectations, with over 50% feeling that their number of publications was not sufficient for their career stage. Productivity was not found to be affected by marital status or parenthood. Inadequacy of time, funding, mentorship, training, and collaboration have been described as barriers to academic productivity [30].

The intention to leave academic medicine is associated with the belief that high quality teaching and clinical care are not rewarded with promotion and tenure as well as the perceived lack of services and facilities at AMC's to provide excellent care [31]. Motivating factors differ between clinician educators and clinician investigators. According to a study of internal medicine faculty, clinician investigators were more likely than clinician educators to be motivated by self expression factors such as being innovative, creative, and free from supervision as well as the attainment of academic achievement [32]. Faculty members who spend more than 50% of time in clinical care have slower career progress, greater dissatisfaction with academic medicine, and less of a commitment to remain in academic medicine [33]. These clinical faculty are at particular risk of attrition if attention is not given to the aspects of academic medicine that they find satisfying.

A survey of faculty who left one medical school found that the most common reasons for leaving included career advancement, low salary, and departmental leadership issues. Men were more likely than women to report leaving due to retirement while women were more likely to leave due to chairman/departmental leadership issues and personal reasons. Women and minority faculty were more likely to be at a lower academic rank when they left [9].

In 2006 the AAMC held a number of focus groups on faculty satisfaction which revealed significant discrepancies between the mission of teaching and research and the administrative and financial pressures pursuant to the bottom line. Aspects of academic practice that were noted to have a positive effect on satisfaction included being valued by patients and colleagues, having a collegial working environment, collaboration and flexibility, especially for research faculty. Negative aspects included not being paid to teach, a negative culture of value by administration, disconnect between faculty and administration, and a lack of transparency [27]. This can be related to the effects of administrative burdens and lack of autonomy that have had a negative impact on satisfaction with the field of medicine in general. It is revealing that from 1970 to 2004 there was a 2,753% increase in the number of hospital administrators compared to a 156% increase in the number of physicians [10]. Dissatisfaction stems from the feeling that education and research, which are the aspects of academic practice that most contribute to job satisfaction, do not generate money, and therefore are not given adequate time or importance by administration:

> And [the administration] really does not value education for the residents or the medical students. That is just not taken into account on what you are doing, how good of a teacher you are. It is basically the bottom line. . .the money that you can bring is what they are valuing. And this being a medical school – that is not the reason we came here [27].

Academic physicians are not primarily motivated by money, but see that money is how they are rated and valued by administration, and this causes dissatisfaction due to a lack of "mission concordance". Although obstetric delivery is the number one CPT code registered in America, from the perspective of hospital administrators, obstetrics does not make money [10]. This discrepancy results in a further disconnect between the goals of the academic obstetrician and that of administration.

8.5 Shifting Gender Balance

It is impossible to address the current practice of obstetrics and gynecology without discussing the issue of gender. With 47% of medical school graduates being female in 2005 compared to 7% in 1966, the entire field of medicine is being affected by this change in the gender demographic. However, this demographic change is not uniformly distributed across specialties. Seventy-eight percent of obstetric/gynecologic residents are now female, a higher percentage than any other specialty. The fields with the next highest percentages of women are pediatrics (73%), dermatology (63%) and psychiatry (55%). Other surgical specialties have a significantly lower representation of women, ranging from 13.9% for thoracic surgery to 42.6% for ophthalmology. The percentage of women in general surgery residencies is 32% [34].

The influx of women into the practice of obstetrics and gynecology should not come as a surprise. In fact, historically, it is the utilization of male providers for women's healthcare that is a relatively new phenomenon. Traditionally "birth attendants" have been other women and the introduction of the male midwife and then the male obstetrician into the birthing chamber was initially met with resistance, and in some societies still is. Obstetrics and gynecology is traditionally a surgical specialty with a difficult lifestyle, characteristics that have been considered a deterrent to women. However, anyone who is under the misimpression that the struggle between family and professional obligations is unique to the twenty-first century should consider reading Laurel Thatcher Ulrich's *A Midwife's Tale: The Life of Martha Ballard, Based on her Diary, 1785–1812.* Whether it is due to patient demand or physician choice, obstetric/gynecologic residency programs have become increasingly female dominated. It is therefore of particular import that the field of obstetrics and gynecology pay attention to the unique needs and values of female physicians, who will represent the majority of ACOG fellows in the near future.

A survey of physicians to explore predictors of satisfaction, work-life balance, and burnout found that women are significantly more satisfied with their careers than men (79 vs. 76%, $p < 0.01$) and that both women and men reported similarly moderate levels of satisfaction with work-life balance (48 vs. 49%), emotional resistance (51 vs. 53%) and high levels of personal accomplishment (74%) [20]. Women also work fewer hours per week (54 vs. 59, $p < 0.05$) and fewer on call hours (41 vs. 52, $p < 0.01$) [20]. The number of children at home has a similar impact on both male and female physicians, and is associated with a decrease in work-life balance and an increase in exhaustion, but not with career satisfaction or personal accomplishment [20].

Gender differences in the satisfaction and retention of academic faculty have been described. In a survey of AMC faculty, 39% of women compared to 66% of men agreed with the statement, "At my medical school female and male faculty members have equal opportunities to be promoted in rank" [35]. Female faculty members are less likely to be promoted to associate or full professor, even after adjustment for work hours and academic productivity. This finding is not accounted for by differential attrition from academic medicine between men and women [36]. Adjusting for professional characteristics and achievements, female medical school faculties neither advance as rapidly nor are compensated as well as similar male colleagues [37, 38]. In AMCs, women have not yet become uniformly distributed across academic rank. In 2004–2005, although women represented 50% of medical school applicants, they represented 42% of residents and fellows, 32% of medical faculty, 15% of full professors and 10% of medical school deans [39]. Gender differences in salary persist after adjustment for number of peer-reviewed publications and clinical revenues [40]. Salary discrepancies by gender increase with increasing seniority [38]. Similar findings in other fields suggest underrepresentation of women in all three branches of federal government, law firm partners, and CEO's of Fortune 500 companies [41]. Possible explanations for these findings include that there is a natural lag between women entering these professions and their rising to the top leadership roles, that there is a glass ceiling that keeps women from obtaining such positions due to their gender, that child-bearing gets in the way, or that women are opting out of career tracts for reasons of dissatisfaction [41].

There is a greater attrition for female than male medical school faculty. According to the AAMC Faculty Roster of years 1981–1997, the percentage of faculty leaving academic medicine in any 10-year period fluctuated from 40 to 47% for women and from 36 to 39% for men [13]. In response to an AAMC survey asking if their medical school cultivates a supportive climate, only 36% (39% of men and 33% of women) agreed [35]. Among academic surgeons at the assistant professor level, 29% of female versus 5% of male surgeons express an interest in leaving academic medicine, and this group of women rated inadequate mentoring as the most important reason for leaving, followed by inadequate career advancement [30]. Female academic surgeons are more likely than their male counterparts to report inadequate mentoring and training for publication. They are also more likely to feel that career advancement opportunities are not equally available to them and to feel a sense of isolation from colleagues [30].

Female academic surgeons are less likely than males to attain high academic positions, become tenured, or publish. However, male and female academic surgeons also report different ultimate career goals with 25% of men compared to 5% of women aspiring to chair a department [30]. In a survey of internal medicine faculty regarding motivating factors, female physicians were significantly more motivated by helping others. Male respondents rated extrinsic rewards such as income, social status, and leadership higher than female respondents, but differences were not statistically significant [32]. While female faculty are less likely to progress in academic rank than male faculty, this may be partially due to differences in personal values regarding career success. One study showed that while both male and female faculty valued aspects of their careers that contributed to the local mission of their school of medicine (quality patient care, relationships with patients and peers), female faculty were less likely to value accomplishments such as national visibility ($p = 0.02$), scholarship ($p = 0.0002$) and leadership ($p < 0.0001$). When asked whose recognition of their work they valued, female physicians were more likely to mention patients, trainees, and local peers than national peers. Female physicians were also less likely to value receiving tenure as a measure of their achievement [33]. Other surveys have found that female medical students and faculty were more interested in patient care and less interested in research than their male colleagues [42]. Since national recognition and scholarship are generally necessary for tenure and promotion, such differences may partially account for differences in men and women in obtaining these academic achievements.

A survey of faculty at the University of Minnesota School of Medicine found that 63% of respondents were satisfied overall, and that there was no difference in overall satisfaction between men and women. However, there were significant differences between men and women in the nuances of satisfaction as reflected by differing responses to subcategories of questions relating to organizational climate and family responsibilities. Despite putting in an equal number of hours at work, women spent an average of 31 h per week on family and household responsibilities compared to 19 h/week for men. Women were more likely to not have a partner (19 vs. 5%) or to have a partner who worked full time (70 vs. 36%). Significantly more women were interested in a part time tenure track (33 vs. 14%). More women than men indicated that obstacles to career success and satisfaction included lack of a part time track, meetings after 5 PM and on weekends, absence of on-site child care and adult care, inadequate formal parental leave policy and inadequate opportunities to pursue research. Men were more likely than women to have a clear picture of where they wanted to be in their academic career in 5–7 years and to have well-defined plans for achieving their career goals [43].

There are important differences in the personal and family lives of male and female physicians. Female academic surgeons are more likely than males to have never married (23 vs. 4%) or to live alone (24 vs. 4%) [30]. Female surgeons are less likely to have children and more likely to delay childbearing [30]. Among married female academic surgeons, 28% are married to another surgeon, 14% to a physician in another specialty, and 90% live in dual career households. In contrast, only 15% of male academic surgeons are married to another physician and 40% live in dual career households. These differences are reflected in a drastic difference in how childcare responsibility is handled in the households of male versus female surgeons. The spouse is primarily responsible for childcare in 63–79% of households of male surgeons compared to 5–29% of households of female surgeons. In contrast, an employed person provides childcare in 40–88% of households of female surgeons compared to 10–29% of households of male surgeons [30, 44]. Female surgeons are more likely than male surgeons to state that both maternity and paternity leave are important and to agree that surgeons should have more part time work opportunities [44].

A study of otolaryngologists reveals parallel findings. Women in this specialty are more likely than men to be divorced or separated, spend more time on household management, and earn 15–20% less than men, even when controlling for hours worked and years since residency ($p < 0.001$) [45]. Male otolaryngologists were more likely than females to report that their spouse or partner provides childcare when a child has a minor illness (89.4% men vs. 14.1% women). Among otolaryngologists, even if the spouse was a physician or surgeon, women were more likely to believe that their spouse expected his career advancement to take priority over their own [45]. The number of hours worked by female nonacademic otolaryngologists declined in proportion to the number of children in their household ($p < 0.001$), while the number of hours worked by male otolaryngologists is independent of the number of children. Female otolaryngologists were more likely then men to report both that their career compromised their family or personal life and that their personal lives interfered with their career [45].

There is no comparable report in the literature that describes the marriage and family experiences of male versus female obstetrician/gynecologists. While one might imagine similarities to the above numbers due to the surgical and lifestyle components of this specialty, it is also possible that the increasing proportion of female obstetrician/gynecologists would lead to significant differences. The increasing number of women who have joined the physician work force while continuing to be involved in traditional roles of childcare and homemaking has been a driving force behind flexible work schedules [46]. The field of obstetrics/gynecology is particularly affected by this change, and how it responds to this challenge may well determine the future of the specialty in general and the future of academic obstetrics/gynecology in particular.

8.6 A New Generation of Physicians

At the same time that the practice of medicine is being confronted with changes demanded by increasing numbers of female physicians, it is also being confronted by the influx of a new generation in which both men and women have very different goals and expectations from their predecessors. The perception of a controllable lifestyle is becoming increasingly important to medical students in their choice of specialty. From 1996 to 2002 there has been a significant linear increase in first-choice ranking of specialties with controllable lifestyle. After controlling for income, work hours, and years of training, having a controllable lifestyle accounted for 55% of the variability in choice of specialty over this time frame ($p < 0.001$) [12]. The increased preference for specialties with a controllable lifestyle has been observed among both male and female medical students, and is not explained solely by the specialty choices of the increasing proportion of female medical students [47]. According to a 2006 AAMC survey of physicians less than 50 years of age, time for family and personal life was rated very important to 66% of men and 82% of women. Long term income potential was very important to 45% of men and 36% of women while the opportunity to advance professionally was very important to 29% of men and 27% of women [48].

The field of medicine in general, and obstetrics and gynecology in particular, must adapt to the changing needs, expectations, and priorities of a new generation. Sixty-five percent of obstetricians are less than 50 years of age [11]. In the words of one ob/gyn baby-boomer, "It seems to me that I happily made choices to develop my career by sacrificing social activities, family, responsibility, and sleep, just to mention a few. When Generation Xs are asked to make these or other personal sacrifices they traditionally are unaccommodating" [10]. The "baby boomer" generation of physicians trained and began practice in what was considered the golden age of medicine. This "Boomer physician" is more likely to be male, work long hours, and see professional life as the overall driving force of daily existence [49]. They worked hard and sacrificed much, and apparently were happy to do so (or at least they did not complain). Residents literally resided in the hospital and, according to the old adage; the only disadvantage to being on call every other night was the missed educational opportunities on the off night. For this dedication the older physicians were granted what has been perceived as a god-like status of unlimited autonomy, great prestige, and luxurious income.

There is little point in applying the chicken versus egg argument to medicine's fall from grace: the bottom line is that physicians that are being trained today are not expected to act like or to be treated like deities. They are conditioned to the work hour regulations set by the Accreditation Council for Graduate Medical Education (ACGME) and they are encouraged to report their programs if infractions are forced on them. They have accepted that they must navigate the insurance industry's web of referrals and pre-approvals before requesting consultations, ordering imaging studies, or scheduling surgeries. They know to check the insurer's preferred formulary before ordering certain medications. They are familiar with the 24-h chart check by which they will be cited for illegible signatures, untimed notes, or the use of abbreviations that have been banned by the Joint Commission. As described by one observer, "In the not-too-distant future, every working physician will for the first time in the United States, have been enveloped in a career long bureaucratic system that systematically leads to a more impersonal, less satisfying work experience" [46]. This contributes to the observation that physicians are increasingly turning to activities outside work for satisfaction. With an increase focus on lifestyle and a decrease in autonomy, it has been observed that "instead of a profession, many now consider this as a job" [10]. If this is so, students entering medicine today are unapologetic about it. The decision to choose specialties based on lifestyle has earned the nickname of the "E-ROAD" – an acronym for Emergency Medicine, Radiology, Ophthalmology, Anesthesiology, and Dermatology [47]. In a medical student perspective published in the Archives of General Surgery in 2002 regarding the lifestyle deterrents to a surgical career the question was aptly put, "Do you live to work or work to live?" [50].

This shift is not unique to the United States. Australia's health system is also under pressure, where it has been observed that junior doctors are making career choices based on flexibility. This has been linked to a growing interest in locum positions, which offer a competitive income along with job flexibility and autonomy, over traditional models in which a strong sense of vocation drives an individual to selflessly work long hours in substandard conditions [51]. The Netherlands have also noted an increasing interest in reduced work hours among young male and female physicians and an interest in less traditional career paths to achieve a fit between actual and preferred work hours [52]. Nor is this interest in lifestyle not unique to medicine. It has also been observed in the fields of business, engineering and the military [12]. Compared to previous generations, generation X is the first in which both parents were likely to work outside the home and are more likely to have divorced parents, leading this generation to seek a greater sense of family and to be less likely to put their jobs first, less willing to sacrifice, and less fixated on titles and the corner office [53].

In order to recruit this generation of physicians, academic medicine must realize that their buyer's market is drawing to an end and that they must focus on mentoring relationships, work-life balance, and career and leadership development [53]. Among young physicians, even the best residents and fellows are not necessarily committed to academic careers. Disincentives include the steep career trajectory of an academic path, the lower income compared to private practice in many specialties, and the lack of mentors or role-models to whom the young generation relates [53].

8.7 Future Directions

It is not necessary to change. Survival is not mandatory.
~W. Edwards Deming

Based on our review of the literature, it becomes clear that the satisfaction and retention of faculty in obstetrics/gynecology requires reform in three inter-related areas:

1. career flexibility
2. patient safety
3. physician wellness.

Without reform in these areas we will continue to confront dissatisfied faculty who struggle with the impossible task of neglecting their personal lives for long clinical hours and the endeavor to produce adequate publications for promotion. The resultant disillusionment and exhaustion is compounded by the feeling of faculty that their educational efforts are undervalued and that they will never measure up to the ideals of themselves or their patients. Restructuring of the practice of obstetrics/gynecology presents an opportunity for the specialty to create a culture in which patient safety is our core principle by identifying and implementing best practices and thereby minimizing professional liability by proactive rather than reactive measures. Such an emphasis on patient safety will "bring us back to what motivated us to go into obstetrics and gynecology – to help women get well and stay well" [54]. Imagine if in so-doing, we could also improve our own wellness!

8.7.1 Career Flexibility

The future of obstetrics/gynecology will demand more flexibility in work hours and in career tracks, both in private practice and in academia. The new breed of obstetrician/gynecologist, who we suggest be called *GEN3X* (generation X and gender XX) will not work the 24-36-48-72-h shifts upon which the current model of ob-gyn practice is based. A survey of obstetrician/gynecologists in Houston found that, in contrast to the limited work hours for residents mandated by the ACGME, most practicing obstetrician/gynecologists worked >80 h per week and that physicians who were men or were >50 years old were more likely to work >80 h/week [55]. The authors concluded that, "resident work-hour restrictions. . .do not reflect the reality of work hours for practicing obstetrician/gynecologists in an urban center" [55]. However, this survey addresses only the status quo, and ignores the impending winds of change. Both the increasing numbers of women in obstetrics/gynecology and the importance placed on lifestyle by the GEN3X physicians have resulted in increased interest in part time practice. By refusing to confront this reality we are more likely to lose talented GEN3X physicians all together than to browbeat them into conforming to the role of the traditional full time physician.

Of full time obstetricians under the age of 50, one in four reports being interested in working part time but not having the option. Among full time obstetricians under the age of 50, those who work less than 44 h per week are more satisfied [11]. Those who are interested in practicing part time but do not have the option are less satisfied than those who do have the option to work part time but were not interested. Again, for obstetricians under the age of 50, higher satisfaction scores are associated with a satisfactory work-personal life balance and with the ability to control work hours and schedules [11]. In addition to enhancing recruitment and retention of GEN3X physicians, the option for part time practice may also delay the retirement of physicians who are later in their career. One in three full time obstetricians over the age of 50 report that they have an interest in working part time, but that they do not have that option [11].

McMurray and colleagues found that part time physicians in the United States are better able to control their work hours and note significantly less stress than full time physicians [56]. Furthermore, they are significantly more satisfied than full time physicians with patient care issues, personal time, administrative issues, and with their jobs overall [56]. Mechaber and collegues found that, in survey of general internists and family physicians, part time physicians were more satisfied, had significantly less burnout, and had more work control than full time physicians. They found no significant difference in job stress or intent to leave practice and no difference between patient satisfaction or trust in their physicians between patients of part time or full time physicians [57]. At a time when trainees are placing a growing emphasis on work-family balance, a core group of satisfied part time physicians could also improve recruitment of medical students [56].

Academic medicine must find a way to reconcile the congruence of career-building and childbearing years. When forced to choose between job and family, where the previous generation may have chosen job, the new generation is likely to choose family. Academic departments will have better recruitment if they do not force their faculty to make this choice. This may mean more flexibility and less than full time options, job sharing, and unpaid leave [53]. If academic medicine does not find a way to accommodate part time or flexible schedules, than it risks losing the skill and labor of the growing number of physicians who are interested in part time employment, many of whom will opt out of academics if the part time option is not available, or if the barriers to such an option are too high [58]. Potential ways in which the AMC can help faculty to balance work and family demands include creation of a part time tenure track, on site and emergency child care and adult care, formal parental leave policies and restructured meeting times. Attention should also be given to addressing both blatant and subtle gender bias and to improved mentoring and faculty development [43].

Other specialties are making significant efforts to improve opportunities for part time physicians. The highest proportion of part time physicians is found in general pediatrics [56]. A recent survey of fellows of the American Academy of Pediatrics shows that the percent of pediatricians working part time has increased from 15% in 2000 to 20% in 2003 to 23% in 2006. An increase was noted among men and women, older and younger physicians, general pediatricians and subspecialists, and those working in inner cities and suburban areas [59]. A survey of the attitudes of pediatric faculty members toward part time positions showed that although 59% of respondents believed that part time faculty were perceived as being less committed to the institution, 69% believed that part time faculty should be eligible for all academic tracks and 73% believed that they should be allowed extension of time to obtain tenure. Furthermore, most respondents (78%) believed that policy changes to support part time faculty would enhance diversity and improve recruitment, retention, and promotion of female faculty [60].

A survey of chairs of departments of pediatrics, medicine, family medicine, and surgery conducted in 1996 found that 85% of departments employed part time faculty, including 94% of pediatric departments, 89% of medicine departments, 86% of family medicine departments, and 72% of surgery departments. Most chairs were very or extremely satisfied with having part time faculty. Advantages included keeping talented people in the workforce who might otherwise leave, leveraging financial resources, and the skills of part time faculty. Disadvantages included a loss of academic productivity, and a lack of shared goals of part time faculty with the department [61].

In 2007 the Association of Specialty Professors, within the Alliance for Academic Internal Medicine convened a task force to establish guidelines for effectively incorporating part time faculty into departments of internal medicine. Part time faculty were defined as those faculty who worked less than full time but whose full professional effort was devoted to the academic institution. Recommendations included: increasing support for work family balance, allowing flexible time as well as part time employment, directly addressing negative perceptions of part-time faculty, developing flexible policies for academic advancement, considering part time faculty for leadership positions and for research awards, and increasing research into best practices for incorporating part time faculty [62]. Challenges addressed by the task force included costs such as fixed overhead and professional liability coverage, promotional clocks and tracks, mentoring difficulties and disruptions in practice coverage [62].

The Society of Internal Medicine has developed a program, the Horn Scholars Program, which is intended to foster a new career track for physicians centering on successful balance of career, family, and social responsibilities [58]. This is a 3 year career development program for clinician educators in internal medicine wishing to work one half of a full time academic position, including care for the medically indigent, while focusing the remainder of the work week caring for dependents. Qualitative interviews of participants in this program (all women ages 30–45) and their division chiefs (all men ages 50–62) were undertaken to explore the positive and negative aspects of part time work for both the faculty participants and the division. Noted advantages of part time work included more research time, more time for family and self, retention of high-quality faculty, and the ability to role model work-family balance. Disadvantages included a slower career trajectory, loss of pay and benefits, and being less than fully integrated into the division. Part time work was viewed fundamentally as either working "less" or working "differently". If one holds the traditional view of a physician as productive, indefatigable and selflessly dedicated, than the focus on part time work is what the part time physician is not doing, rather than what they are contributing. If a medical career is considered in the context of long-term values and priorities, then the part time physician can become valued as a skilled and dedicated member of the faculty. It is such a shift in cultural attitudes that must take place in medicine in general, and in academic medicine in particular, if the contribution of the part time faculty member is truly to be embraced, and that valuable physicians who might otherwise opt out of academic medicine might be retained [58].

One argument against the many calls in support of increased opportunities for part time practice is that such an approach fails to address the ever increasing and unsustainable work load of full time academic faculty, typically requiring 60–80 h or more per week in order to meet productivity expectations [63]. In a commentary on this subject, Dr. Deborah Helitzer suggests that the work-life balance of all academic faculty members could be improved by adaptations such as reduced bureaucratic requirements, relaxation of promotion time lines, accommodations of family needs with on site services such as child-care, adult-care and dry-cleaning, and student loan forgiveness programs for those pursuing academic careers [63]. This highlights the perspective that the key to the future practice of medicine is not so much the number of hours worked, but in allowing more career flexibility for all physicians, at varying stages of their careers. Reforms for the field of obstetrics and gynecology as proposed by Weinstein and Wolfe have included limiting full time work hours to 50 h/week, use of laborists for obstetric coverage, part time liability policies, job sharing, and education in time and money management [64].

The call for increased flexibility in order to improve academic recruitment has been made in other countries. In the United Kingdom, the lack of enthusiasm for an academic career has been the focus of a Modernizing Medical Careers initiative which has called for increased flexibility, particularly in fields such as pediatrics due to the high proportion of women. Proposed reforms have included the use of competency assessment rather than "time served" to shorten academic training, flexibility in moving between clinical and academic paths, creation of positions with less on-call load, job sharing, and equal recognition for research and teaching as core academic activities [65].

8.7.2 Patient Safety

In the midst of increasing requirements from insurance companies and regulatory bodies, the need to balance budgets in the face of decreasing reimbursement, and the controversy surrounding professional liability, it is easy to forget that the physician should be the greatest advocate for patient safety. Nowhere should this be truer than in the practice of obstetrics/gynecology, where inherent health disparities and unpredictable clinical outcomes combine to increase the vulnerability of patients and to thwart the best intentions of the most dedicated of physicians. Part of physician dissatisfaction may come from the fact that we really do want to provide safe care, and it seems that we are working in a system that does not allow us to live up to the expectations of ourselves or our patients. The high professional liability costs and so-called "malpractice crisis" combined with high physician dissatisfaction have been described as creating a perfect storm for the specialty of obstetrics/gynecology. In order to weather this storm, we must create a culture of patient safety [54]. Strategies to improve patient safety include adopting a transparent systems based approach to errors and near misses, improving communication among providers and between patients and providers, and developing a patient-centered interdisciplinary approach to patient care [54]. Making patient safety a fundamental principle of our practice can allow us to regain control that has been lost to lawyers and policy makers, and to regain the motivation to practice obstetrics/gynecology – "to help women get well and stay well" [54].

One of the key concepts in the patient safety movement is the acceptance that health care providers, as human beings, are fallible and will make mistakes, but that it is the responsibility of the system to have safeguards in place that prevent mistakes from jeopardizing patient safety. A useful analogy likens human errors to holes in Swiss cheese. A poor system, like Emmenthal cheese, has large holes that are easily penetrated by system failures. A safe system, like Havarti cheese, has tighter defenses that are less likely to be penetrated by system failures [66]. Veltman has done an excellent job of pointing out the "holes" that are inherent in the current practice of obstetrics/gynecology, most of which are familiar to anyone within the specialty [66]. These include:

1. the frequent need for the obstetrician to be simultaneously responsible for patients in the office, in labor and delivery, and in the operating room
2. high patient volume
3. poor sign out practices
4. inadequate protocols
5. acquiescence to unsafe patient requests
6. off-site monitoring of high-risk situations
7. hierarchy at the expense of teamwork
8. inadequate back-up
9. failure to recognize the effects of "human factors" such as fatigue [66].

Veltman goes on to describe strategies for tightening the system defenses of obstetric practice, or "getting to Havarti." Potential methods include structured communication, simulation training, use of protocols, increased in-house presence, attention to high-risk issues, full "24/7" labor and delivery staffing, and the use of data to support practice-based learning. It is also necessary to have a departmental infrastructure that includes effective peer review, multi-disciplinary development of protocols and review of near-misses, and ongoing patient safety and risk management education [66].

Quality improvement can have a positive impact on physician satisfaction. The presence of quality problems in a practice has been found to be independently associated with increased professional isolation, work-life stress, and practice dissatisfaction [17]. Although loss of autonomy has previously been associated with a decrease in physician satisfaction, one survey found that involvement in quality improvement activities is associated with a positive trend in satisfaction [17]. Rather than allowing ourselves to be the victims of frivolous lawsuits and petty regulations, we should become the leading champions of patient safety and quality of care. Obstetrician/gynecologists who have a greater interest in continuing medical education have a significantly higher satisfaction rating than those who considered such programs to be burdensome [23]. Involvement in quality improvement is paramount to being a member of the medical profession, as a profession is defined not just by doing what it does well, but by doing what it does better all the time [67]. This requires continuous quality improvement and experiential learning that improves behavior as well as knowledge and the use of outcomes data to continuously raise performance levels [67]. Data is increasing regarding the efficacy of such methods, and we as physicians, must make an important choice whether to embrace or to resist incorporating them into our medical practice.

The current medical system demands that physicians, especially obstetricians, work long and unpredictable hours. Patient safety concerns have forced restrictions on the number of hours worked by residents but not by attending physicians. Allowing for shorter, more flexible work hours is necessary not only for physician satisfaction, but more importantly for patient safety. Although there is little clinical data linking physician fatigue to patient outcomes, the potential effects of fatigue on patient safety cannot continue to be ignored. Medicine is a human endeavor, and the human need for sleep has been well studied. The National Sleep Foundation recommends 8 h of sleep per night for an adult. Sleep deprivation results from both interrupted sleep and insufficient sleep. Sleep cannot be stored up, and recovery from a period of insufficient sleep requires at least 2–3 full nights of adequate uninterrupted sleep [68].

Other industries have recognized the effects of sleep deprivation. The National Transportation Safety Board rates excessive sleepiness as the second leading cause of driving accidents in the United States and the Federal Railroad Administration showed that fatigue has played a role in 25% of train accidents [68]. Physicians are fallible human beings who are prone to mistakes and who frankly, just like everyone else, need sleep. A recent retrospective study of procedures performed by attending surgeons and obstetrician/gynecologists demonstrated an increased rate of complications in procedures that were performed following a night of call that allowed for less than 6 h of sleep opportunity [69]. The counter-arguments have been loudly stated: shorter work hours have not been proven to improve patient outcomes; there are benefits to continuity of care; and there are critical patient safety issues that are inherent in increasing the number of hand-offs. However, these are NOT excuses for keeping the status quo. As stated in the ACOG committee opinion on fatigue and safety, "The inability to document improved outcome after residency work-hour reform does not mean that fatigue and safety can be ignored by practicing physicians" [68]. Rather, we need to standardize care plans and revamp the sign-out processes so that patient care can be safely transferred from one physician to the next without interrupting the continuity of care. Promising options include quality of care bundles such as those that have reduced ventilator associated pneumonia and the use of standardized regimens, such as an oxytocin protocol as proposed by Hayes and Weinstein [70, 71].

In order to implement many of these system changes to improve patient safety, the cost of professional liability insurance must be controlled, and the option for part time policies must become widespread [4]. In order to improve both patient safety and physician lifestyle, each physician must work shorter shifts without conflicting responsibilities, necessitating more physicians for a given obstetric practice. On the other hand, the expense of professional liability policies creates an incentive to try to cover the obstetric service with fewer physicians, in order to decrease the number of expensive policies that a given obstetric practice must purchase. This leads to fewer doctors working longer hours with more competing responsibilities, intuitively causing patients to be less safe and physicians to be less satisfied. Still, according to ACOG, "Even though there may be some economic impact of changing schedules to accommodate avoidance of fatigue, patient care and safety must take priority over economic concerns" [68].

The Laborist model proposed by Weinstein is one unique way of reconciling both patient safety and physician lifestyle [72]. Employing an obstetrician to work 10–14 h shifts with a maximum of 42 h/week covering only the labor floor allows for predictable and controllable work hours, which has been associated with improved patient satisfaction. Physicians wanting to work fewer hours for personal reasons may work a limited number of these shifts which would give them the flexibility to continue to practice rather than quit obstetrics or leave practice altogether. It would also help to close many of the "holes in the cheese" described above. This may also reduce the contribution to dissatisfaction that comes from not being able to meet too many competing expectations, and improve the satisfaction that comes from the inherent opportunity to participate in quality improvement. Variations on the laborist theme have been initiated at many institutions since this model was first proposed. Studies of patient safety and physician satisfaction in these settings should be undertaken to determine how this can influence the field as a whole.

8.7.3 Physician Wellness

The patient safety and quality of care movements rely on certain "indicators" of quality that health care systems can measure, report, and improve. A revolutionary concept proposed by Jean Wallace and colleagues in the November 2009 edition of the Lancet is that one such quality indicator should be physician wellness [73]. Should physician wellness be an indicator of quality? Why should the health care system care about the wellness of physicians? Isn't a physician simply a "cog in the wheel" of the healthcare system, an entity obligated to produce desired patient outcomes amid diminishing reimbursement, external cost controls, and endless rules and regulations? There is data to suggest otherwise. Physicians' overall job satisfaction has been found to have a positive effect on patients' adherence to treatment [74]. Both medical students and physicians who have poor personal health profiles are less likely than those who are healthy to recommend evidence-based screening to their patients or to provide counseling for a healthy lifestyle [73]. Physician burnout is associated with reduced productivity and efficiency [73]. Physician work-related stress has been connected to reduced standards of care such as taking shortcuts or not following protocols, that have been linked to diminished safety [73]. Furthermore, it is expensive (at least $150,000–300,000) to replace a physician. Physician wellness is an indicator that is both measurable and actionable [73]. Only if efforts are made to strive for improvement of this indicator can we discover whether interventions improve costs and patient outcomes.

Medical research, conducted at AMCs', has devoted countless hours toward the study of disease. New ways to study patient safety and quality improvement are being developed at a rapid pace. A final crucial component to improving medical care is to focus on the subject of physician wellness. How can it be that a profession that is chock-full of opportunities for intellectual stimulation, interpersonal relationships, achievement of technical skills and a myriad other rewards can be plagued by high levels of dissatisfaction, burnout, depression, substance abuse, and suicide? [73]. Prevalent physician personality traits such as perfectionism, workaholics, and the type A personality have been associated with adverse health outcomes such as burnout, depression, anxiety, eating disorders, and cardiovascular disease [73]. Furthermore, we physicians are not very good at taking care of ourselves. Many do not have a personal doctor, self-prescribe drugs, and neglect to seek help for physical or mental symptoms. Physicians may be discouraged from seeking help due to inherent personalities and a perceived stigma of needing help, but also my be deterred by the fear that treatment, especially for problems related to mental health and substance abuse, can lead to investigation by licensing boards and damage to their careers [73]. Physicians also report working when physically unwell and feeling pressured from patients and colleagues to appear well even when they are sick [73]. In a survey of junior doctors regarding anticipated response to a hypothetical illness, 61% reported that they would "work and wait and see" if they were vomiting all night, 83% if they had blood in their urine, 76% if they suspected a stomach ulcer, and 73% if they had severe anxiety [73].

Such professional hypocrisy – dispensing to others, yet personally ignoring advice about leading a balanced life – is decried in Dr. Charles Hatem's words on "renewal in the practice of medicine" [75]. In order to achieve renewal, Hatem states that we physicians must realize that we are not limitless resources [75]. We need to develop core values around which we can balance our lives. We need to practice self-preservation and remember to take non-working, guilt-free vacations. We need to search for meaning in both our personal and professional lives, while regaining control of our own attitudes and passion for medicine.

A fulfilling career in medicine means not only decreasing those aspects of the profession that are most dissatisfying, but also concentrating on those aspects that are most satisfying [76]. The field of medicine may benefit from data in nonmedical industries that indicate improved measures of success with improved employee satisfaction [75]. This will require further identification of aspects of work where the cost of de-motivation and the rewards of improved motivation can make the most substantial difference, while accepting that in some ways the nature of our work itself may have to change. The extent to which faculty physicians are able to focus on the aspect of work that is most meaningful to them has been found to have a strong inverse relationship to their risk of burnout [77]. Optimization of career fit may promote physician satisfaction and reduce attrition of academic physicians [77]. The question for the future is one of how to fit the practice of academic obstetrics/gynecology with the goals and attitudes of the new GEN3X.

Another proposed solution to faculty attrition, which can be connected to physician wellness, has been the formation of formal mentoring programs. Mentoring in academic medicine has been reported to influence personal development, career guidance, career choice, and research productivity, but there is currently no strong evidence to support this perception [78]. The Junior Faculty Development Program (JDFP) at Penn Sate College of Medicine is an example of a formalized program developed specifically for the purpose of promoting the development and advancement of junior faculty to replace the old "sink-or-swim" approach to faculty development. This program involves completion of an individual project by each participant and follows a course format with a recommended 4 h per week of protected time (indicative of institutional support). Participating junior faculty are asked to approach a mentor, preferably from another department, in order to empower junior faculty to initiate and develop new professional relationships and to facilitate interdepartmental collaboration. After 2 years of the program, nearly all participants reported that they were satisfied with the program and would recommend it to others. They reported perceived increases in abilities related to research review, teaching, understanding promotion, career planning, communication, and grant management [79].

In a pilot study of faculty mentoring at a Harvard Medical School affiliate hospital, both mentors and mentees reported significant benefits from the program. Both reported feeling better supported by the department and feeling greater camaraderie. One hundred percent of mentors reported that the program gave them an opportunity for self-reflection and made them feel more connected. Mentees reported that important benefits of the program included career planning, balancing work and family, knowledge about organizational structure of the hospital and leaderships skills. In selection of a mentor, ability to connect was found to be more important than gender [80]. Done well, mentorship programs may not only improve the retention and recruitment of junior physicians, but they may renew the passion for medicine in the more senior physicians.

At first glance the literature on the subject of physician satisfaction and retention reads as both a lamentation of the bygone glory days of medicine and a harbinger of an impersonal, highly regulated, and algorithm-driven system devoid of any gratification. The public wants to feel safe in the hands of the healthcare system. GEN3X wants to have it all – a great career and a great lifestyle. We need nothing more than determination to create a system that achieves both of these goals. Such a system is an aspiration worthy to be pursued by the field of academic obstetrics/gynecology. Our survival depends on it. The practice of medicine is being transformed, but seen in the right light, this may represent the prospect of the revival of the greatness of this profession: the potential of medical science to understand and treat disease, and the opportunity of one human being to understand and treat another. The winds of change bring with them an opportunity for the creation of a new culture of medicine – one that respects the individual personhood of both the patient and the physician.

References

1. Sheldon, G.F. 2003. Great expectations: The 21st century health workforce. *American Journal of Surgery* 185(1):35–41.
2. Goodman, D.C., and E.S. Fisher. 2008. Physician workforce crisis? Wrong diagnosis, wrong prescription. *The New England Journal of Medicine* 358(16):1658–1661.
3. Cooper, R.A. 2009. Regional variation and the affluence-poverty nexus. *Journal of American Medical Association* 302(10):1113–1114.
4. Iglehart, J.K. 2008. Grassroots activism and the pursuit of an expanded physician supply. *New England Journal of Medicine* 358(16):1741–1749.
5. Kirch, D.G., and D.J. Vernon. 2008. Confronting the complexity of the physician workforce equation. *Journal of American Medical Association* 299(22):2680–2682.

6. Nasca, T.J., R.L. Phillips, K. Reed, R. Rockwell, and V.L. Seltzer. 2010. Nineteenth report: Enhancing flexibility in graduate medical education. http://www.cogme.gov/19thReport/default.htm. Accessed 13 Jan 2010.
7. Association of American Medical Colleges. 2010. Medical school enrolment plans through 2013: Analysis of the 2008 AAMC survey. <http://www.aamc.org/workforce/enrollment/enrollmentreport>. Accessed 13 Jan 2010.
8. Rittenhouse, D.R., E. Mertz, D. Keane, and K. Grumbach. 2004. No exit: An evaluation of measures of physician attrition. *BMC Health Services Research* 39(5):1571–1588.
9. Cropsey, K.L., S.W. Masho, R. Shiang et al. 2008. Why do faculty leave? Reasons for attrition of women and minority faculty from a medical school: Four-year results. *Journal of Womens Health (Larchmt)* 17(7):1111–1118.
10. Cowan, B.D. 2009. Women's health care vital signs: Stable or not? *American Journal of Obstetrics and Gynecology* 200(3):325–328.
11. Anderson, B.L., R.W. Hale, E. Salsberg, and J. Schulkin. 2008. Outlook for the future of the obstetrician-gynecologist workforce. *American Journal of Obstetrics and Gynecology* 199(1):88.e1–88.e8.
12. Dorsey, E.R., D. Jarjoura, and G.W. Rutecki. 2003. Influence of controllable lifestyle on recent trends in specialty choice by US medical students. *Journal of American Medical Association* 290(9):1173–1178.
13. Hershel, A, and J. Lang. 2008. The long-term retention and attrition of U.S. medical school faculty. *Analysis in Brief* 8(4), June 2008.
14. Autry, A., D. Irby, and C. Hodgson. 2007. Faculty attrition in obstetrics and gynecology. *American Journal of Obstetrics and Gynecology* 196(6):603.e1–4; discussion 603.e5.
15. Zuger, A. 2004. Dissatisfaction with medical practice. *New England Journal of Medicine* 350(1):69–75.
16. Holsinger, J.W., and B. Beaton. 2006. Physician professionalism for a new century. *Clinical Anatomy* 19:473–479.
17. Quinn, M.A., A. Wilcox, E.J. Orav, D.W. Bates, and S.R. Simon. 2009. The relationship between perceived practice quality and quality improvement activities and physician practice dissatisfaction, professional isolation, and work-life stress. *Medical Care* 47(8):924–928.
18. Solomon, J. 2008. How strategies for managing patient visit time affect physician job satisfaction: A qualitative analysis. *Journal of General Internal Medicine* 23(6):775–780.
19. Bovier, P.A., and T.V. Perneger. 2003. Predictors of work satisfaction among physicians. *European Journal of Public Health* 13(4):299–305.
20. Keeton, K., D.E. Fenner, T.R., Johnson, and R.A. Hayward. 2007. Predictors of physician career satisfaction, work-life balance, and burnout. *Obstetrics & Gynecology* 109(4):949–955.
21. Landon, B.E., J.D. Reschovsky, H.H. Pham, and D. Blumenthal. 2006. Leaving medicine: The consequences of physician dissatisfaction. *Medical Care* 44(3):234–242.
22. Leigh, J.P., D.J. Tancredi, and R.L. Kravitz. 2009. Physician career satisfaction within specialties. *BMC Health Services Research* 9:166.
23. Bettes, B.A., A.L. Strunk, V.H. Coleman, and J. Schulkin. 2004. Professional liability and other career pressures: Impact on obstetrician-gynecologists' career satisfaction. *Obstetrics & Gynecology* 103(5 Pt 1):967–973.
24. Council on Long Range Planning and Development, American Medical Association. 1987. The future of obstetrics and gynecology. *Journal of American Medical Association* 258(24):3547–3553.
25. Klagholz, J., and A.L. Strunk. 2009. Overview of the 2009 ACOG survey on professional liability. *ACOG Clinical Review* 14(6):1–16.
26. Gabbe, S.G., J. Melville, L. Mandel, and E. Walker. 2002. Burnout in chairs of obstetrics and gynecology: Diagnosis, treatment, and prevention. *American Journal of Obstetrics and Gynecology* 186(4):601–612.
27. Anonymous. 2006. *Medical faculty job satisfaction: Thematic overviews from ten focus groups*. Washington, DC: Association of American Medical Colleges.

28. Schloss, E.P., D.M. Flanagan, C.L. Culler, and A.L. Wright. 2009. Some hidden costs of faculty turnover in clinical departments in one academic medical center. *Academic Medicine* 84(1):32–36.
29. Bell, D.J., J. Bringman, A. Bush, and O.P. Phillips. 2006. Job satisfaction among obstetrician-gynecologists: A comparison between private practice physicians and academic physicians. *American Journal of Obstetrics and Gynecology* 195(5):1474–1478.
30. Schroen, A.T., M.R. Brownstein, and G.F. Sheldon. 2004. Women in academic general surgery. *Academic Medicine* 79(4):310–318.
31. Lowenstein, S.R., G. Fernandez, and L.A. Crane. 2007. Medical school faculty discontent: Prevalence and predictors of intent to leave academic careers. *BMC Medical Education* 7:37.
32. Wright, S.M., and B.W. Beasley. 2004. Motivating factors for academic physicians within departments of medicine. *Mayo Clinic Proceedings* 79(9):1145–1150.
33. Buckley, L.M., K. Sanders, M. Shih, and C.L. Hampton. 2000. Attitudes of clinical faculty about career progress, career success and recognition, and commitment to academic medicine. Results of a survey. *Archives of Internal Medicine* 160(17):2625–2629.
34. Brotherton, S.E., and S.I. Etzel. 2009. Graduate medical education, 2008–2009. *Journal of American Medical Association* 302(12):1357–1372.
35. Bunton, S.A. 2008. Differences in U.S. medical school faculty job satisfaction by gender. *Analysis in Brief* 8(7), November 2008.
36. Tesch, B.J., H.M. Wood, A.L. Helwig, and A.B. Nattinger. 1995. Promotion of women physicians in academic medicine. Glass ceiling or sticky floor? *Journal of American Medical Association* 273(13):1022–1025.
37. Wright, A.L., K. Ryan, P. St Germain, L. Schwindt, R. Sager, and K.L. Reed. 2007. Compensation in academic medicine: Progress toward gender equity. *Journal of General Internal Medicine* 22(10):1398–1402.
38. Ash, A.S., P.L. Carr, R. Goldstein, and R.H. Friedman. 2004. Compensation and advancement of women in academic medicine: Is there equity? *Annals of Internal Medicine* 141(3): 205–212.
39. Anonymous. 2005. *Women in U.S. Academic Medicine: Statistics and Medical School Benchmarking*. Washington, DC: Association of American Medical Colleges.
40. Wright, A.L., L.A. Schwindt, T.L. Bassford et al. 2003. Gender differences in academic advancement: Patterns, causes, and potential solutions in one US college of medicine. *Academic Medicine* 78(5):500–508.
41. Belkin, L. 2003. The Opt-Out revolution. *The New York Times*, 26 Oct 2003.
42. Buckley, L.M., K. Sanders, M. Shih, S. Kallar, and C. Hampton. 2000. Obstacles to promotion? Values of women faculty about career success and recognition. committee on the status of women and minorities, Virginia Commonwealth University, Medical College of Virginia campus. *Academic Medicine* 75(3):283–288.
43. Shollen, S.L., C.J. Bland, D.A. Finstad, and A.L. Taylor. 2009. Organizational climate and family life: How these factors affect the status of women faculty at one medical school. *Academic Medicine* 84(1):87–94.
44. Troppmann, K.M., B.E. Palis, J.E. Goodnight Jr, H.S. Ho, and C. Troppmann. 2009. Women surgeons in the new millennium. *Archives of Surgery* 144(7):635–642.
45. Grandis, J.R., W.E. Gooding, B.A. Zamboni et al. 2004. The gender gap in a surgical subspecialty: Analysis of career and lifestyle factors. *Archives of Otolaryngology-Head & Neck Surgery* 130(6):695–702.
46. Igel, L.H. 2007. The physician condition: The age of economic medicine. *The Physician Executive* 33(1):42–45.
47. Dorsey, E.R., D. Jarjoura, and G.W. Rutecki. 2005. The influence of controllable lifestyle and sex on the specialty choices of graduating U.S. medical students, 1996–2003. *Academic Medicine* 80(9):791–796.
48. AAMC statement on the Physician Workforce, June 2006. http://www.aamc.org/workforce/workforceposition.pdf. Accessed 13 Jan 2010.

49. Weinstein, L. 2008. The unbearable unhappiness of the ObGyn: A crisis looms. *OBG Management* 20(12):34–42.
50. Henningsen, J.A. 2002. Why the numbers are dropping in general surgery: The answer no one wants to hear – Lifestyle! *Archives of Surgery* 137(3):255–256.
51. Skinner, C.A. 2006. Re-inventing medical work and training: A view from generation X. *The Medical Journal of Australia* 185(1):35–36.
52. Heiligers, P.J.M., and L. Hingstman. 2000. Career preferences and the work-family balance in medicine: Gender differences among medical specialists. *Social Science and Medicine* 50(9):1235–1246.
53. Bickel, J., and A.J. Brown. 2005. Generation X: Implications for faculty recruitment and development in academic health centers. *Academic Medicine* 80(3):205–210.
54. Pearlman, M.D., and P.A. Gluck. 2005. Medical liability and patient safety: Setting the proper course. *Obstetrics & Gynecology* 105(5 Pt 1):941–943.
55. Promecene, P.A., K.M. Schneider, and M. Monga. 2003. Work hours for practicing obstetrician-gynecologists: The reality of life after residency. *American Journal of Obstetrics and Gynecology* 189(3):631–633.
56. McMurray, J.E., P.J. Heiligers, R.P. Shugerman et al. 2005. Part-time medical practice: Where is it headed? *American Journal of Medicine* 118(1):87–92.
57. Mechaber, H.F., R.B. Levine, L.B. Manwell et al. 2008. Part-time physicians. . .prevalent, connected, and satisfied. *Journal of General Internal Medicine* 23(3):300–303.
58. Harrison, R.A., and J.L. Gregg. 2009. A time for change: An exploration of attitudes toward part-time work in academia among women internists and their division chiefs. *Academic Medicine* 84(1):80–86.
59. Cull, W.L., K.G. O'Connor, and L.M. Olson. 2010. Part-time work among pediatricians expands. *Pediatrics* 125(1):152–157.
60. Kahn, J.A., S.J. Degen, M.E. Mansour et al. 2005. Pediatric faculty members' attitudes about part-time faculty positions and policies to support part-time faculty: A study at one medical center. *Academic Medicine* 80(10):931–939.
61. Socolar, R.R., and L.S. Kelman. 2002. Part-time faculty in academic pediatrics, medicine, family medicine, and surgery: The views of the chairs. *Ambulatory Pediatric* 2(5):406–413.
62. Linzer M, C. Warde, R.W. Alexander et al. 2009. Part-time careers in academic internal medicine: A report from the association of specialty professors part-time careers task force on behalf of the alliance for academic internal medicine. *Academic Medicine* 84(10):1395–1400.
63. Helitzer, D. 2009. Commentary: Missing the elephant in my office: Recommendations for part-time careers in academic medicine. *Academic Medicine* 84(10):1330–1332.
64. Weinstein, L., and H.M. Wolfe. 2007. The downward spiral of physician satisfaction: An attempt to avert a crisis within the medical profession. *Obstetrics & Gynecology* 109(5): 1181–1183.
65. Winyard, P.J., H.D. Cass, T.J. Stephenson, A.R. Wilkinson, and R.E. Olver. 2006. Developing critical mass and growing our own academics. *Archives of Disease in Childhood* 91(12): 1027–1029.
66. Veltman, L.L. 2007. Getting to havarti: Moving toward patient safety in obstetrics. *Obstetrics & Gynecology* 110(5):1146–1150.
67. Batalden, P., and F. Davidoff. 2007. Teaching quality improvement: The devil is in the details. *Journal of American Medical Association* 298(9):1059–1061.
68. American College of Obstetricians and Gynecologists. 2008. Fatigue and patient safety. ACOG committee opinion no. 398. *Obstetrics & Gynecology* 111(2):471–473.
69. Rothschild, J.M., C.A. Keohane, S. Rogers et al. 2009. Risks of complications by attending physicians after performing nighttime procedures. *Journal of American Medical Association* 302(14):1565–1572.
70. Cocanour, C.S., M. Peninger, B.D. Domonoske et al. 2006. Decreasing ventilator-associated pneumonia in a trauma ICU. *Journal of Trauma – Injury, Infection and Critical Care* 61(1):122–129.

71. Hayes, E.J., and L. Weinstein. 2008. Improving patient safety and uniformity of care by a standardized regimen for the use of oxytocin. *American Journal of Obstetrics and Gynecology* 198(6):622.e1–622.e7.
72. Weinstein, L. 2003. The laborist: A new focus of practice for the obstetrician. *American Journal of Obstetrics and Gynecology* 188(2):310–312.
73. Wallace, J.E., J.B. Lemaire, and W.A. Ghali. 2009. Physician wellness: A missing quality indicator. *Lancet* 374(9702):1714–1721.
74. DiMatteo, M.R., C.D. Sherbourne, R.D. Hays et al. 1993. Physicians' characteristics influence patients' adherence to medical treatment: Results from the medical outcomes study. *Health Psychology* 12(2):93–102.
75. Hatem, C.J. 2006. Renewal in the practice of medicine. *Patient Education and Counseling* 62(3):299–301.
76. Brown, S., and R.B. Gunderman. 2006. Viewpoint: Enhancing the professional fulfillment of physicians. *Academic Medicine* 81(6):577–582.
77. Shanafelt, T.D., C.P. West, J.A. Sloan et al. 2009. Career fit and burnout among academic faculty. *Archives of Internal Medicine* 169(10):990–995.
78. Sambunjak, D., S.E. Straus, and A. Marusic. 2006. Mentoring in academic medicine: A systematic review. *Journal of American Medical Association* 296(9):1103–1115.
79. Thorndyke, L.E., M.E. Gusic, J.H. George, D.A. Quillen, and R.J. Milner. 2006. Empowering junior faculty: Penn state's faculty development and mentoring program. *Academic Medicine* 81(7):668–673.
80. Tracy, E.E., R. Jagsi, R. Starr, and N.J. Tarbell. 2004. Outcomes of a pilot faculty mentoring program. *American Journal of Obstetrics and Gynecology* 191(6):1846–1850.

Chapter 9
Ethics in Academic Medicine

Anne Drapkin Lyerly

Abstract Membership in academic medicine carries with it special challenges and responsibilities in such matters as patient-physician relations, physician conduct and practice, and conflict of interest. This chapter addresses a limited selection of ethical issues in academic medicine that particularly apply to women's healthcare.

Keywords Ethics · Bioethics · Professional ethics · Conflict of interest · Innovative practice · Gender schemas · Conscience

9.1 Introduction

The provision of health care for women is critical both to their own well being, as well as the health of their families, communities, and society broadly. Its effects are both immediate and long term, with the both benefits and shortfalls of care potentially stretching across generations. Yet according to the World Health Organization, societies and their health systems have failed to adequately address the issues that affect women across their lifespan [1]. Thus obstetricians and gynecologists – and those who will educate and train them – face considerable challenges and opportunities in improving women's healthcare worldwide.

Central among these challenges are ethical considerations for an evolving profession. Obstetrics and gynecology is known for generating important – and divisive – ethical debate. Situated as it often is at the beginning (sometimes the end) of life, around health issues such as reproduction that are imbued with meaning, ethical tensions in reproductive medicine abound. Furthermore, rapidly expanding capabilities in diagnosis and treatment as well as enduringly polarizing questions about abortion and contraception generate numerous ethical dilemmas for physicians caring for patients.

A.D. Lyerly (✉)
Department of Social Medicine and Center for Bioethics, University of North Carolina School of Medicine, Chapel Hill, NC 27599, USA
e-mail: alyerly@email.unc.edu

W.F. Rayburn, J. Schulkin (eds.), *Changing Landscape of Academic Women's Health Care in the United States*, International Library of Ethics, Law, and the New Medicine 48,
DOI 10.1007/978-94-007-0931-7_9,

Such profound questions arising in the clinical encounter have demanded much attention of those addressing ethics in obstetrics and gynecology. Yet less attention has been directed toward what might be called questions of "professional" ethics: what ethical considerations should guide obstetrician-gynecologists when they teach? when they encounter economic or intellectual opportunity? when they craft a workplace, a workforce, an organization? The changing landscape of academic obstetrics and gynecology has intensified some longstanding challenges in professional ethics, and generated scores of new ones. While it would be impossible to address in a single chapter the breadth of controversies facing contemporary academic obstetrics and gynecology (much less ethical issues in reproductive medicine more broadly construed), described here are a limited selection of the most pressing of these challenges, organized around a trio of tensions faced by individual physicians, by institutions, and the by profession respectively.

9.2 Ethics in Medicine

Before describing issues specific to academic medicine, it may be helpful to consider the role of ethics in medicine broadly. The last four decades have been marked by a growing recognition of the importance of ethics in medicine, science and health. In the early 1970s, the term "bioethics" was coined, and with it emerged an interdisciplinary effort to formalize a bridging of scientific knowledge with human value systems [2]. Scholarship in bioethics has addressed ever more pressing issues regarding medicine and society, ranging from questions about medicine's appropriate role at beginnings and endings of life to questions about the appropriate and fair distribution of limited resources for health care.

There is no single "best" way to analyze ethical issues in medicine, and the discussion that follows draws on a range of theories and methodologies. Indeed, some have invoked the metaphor of a "toolbox" as a useful way to think about employing different approaches to different ethical problems [3]. In other words, different ethical approaches or "tools" might be appropriate for different issues that arise. For instance, the first section (on Challenges for Academic Physicians) draws heavily on *principle based ethics*. This approach relies on a set of principles or rules for guidance or justification. The major principles commonly invoked as guides to professional action include autonomy, beneficence, nonmaleficence, and justice [4]. Elsewhere invoked is feminist ethics, which uses the tools of feminist theory to examine ethical issues across medicine [5]. This approach can reveal how gendered conceptions affect women as patients and doctors, distort the tools that philosophers and bioethicists use, and identify and challenge dominance and oppression due to gender, race, class or other characteristics. Other approaches include case-base reasoning, communitarian ethics, and virtue based ethics. All may be useful in addressing the issues described here as well as other ethical issues that arise across academic medicine.

9.3 Challenges for Academic Physicians: Ethics of Teaching and Research

Changes in the landscape of academic medicine have intensified challenges faced by faculty in academic obstetrics and gynecology. Demand for financial productivity in the face of escalating costs and decreasing reimbursement has limited time and opportunities for academic pursuits [6], including education and research.

Consider first medical education, and ethical issues faced by *clinician-educators,* whose role has become more central due to increasing clinical and educational needs of expanding departments (see Chapter 4). Effective teaching and training is critical to ensuring that the next generation of obstetrician-gynecologists will provide safe, patient-centered and evidence based care to women in the future. Yet the process of training raises several potentially contentious issues, given a learner's involvement in the care of a sick or parturient patient not only as a student but as a trainee or even a provider of health care. A student or trainee's involvement brings with it potential benefits to the attending, to the patient, and to the student herself. Yet because the primary end of education is pedagogy rather than promoting the best interests of patients (although the latter is certainly a goal), tensions may arise for an educator who must consider her respective roles as educator and healthcare provider. The challenge of managing these potentially conflicting roles is longstanding, but as time for education has become more constrained by increasing demands for revenue and limitations on trainee duty hours [7], tensions between the physician's responsibilities to educate and train and their responsibilities to respect and promote patients' interests have been accentuated.

Over the last several decades, the principle of *respect for autonomy* has emerged as one of the most important concepts in biomedical ethics, and should guide clinician educators in their decisions about involving trainees in patient care. The principle holds that persons should be free to choose and act without controlling constraints imposed by others. Respecting autonomy means to respect a patient's capacities and perspectives, including her right to hold certain views, make certain choices and – of particular relevance to clinician educations – determine whether or not students or trainees may participate in their care [8]. Respect for autonomy in the care of patients in obstetrics and gynecology may be particularly pressing given the private, personal and sometimes pivotal issues around sexuality and childbearing that characterize this area of medicine [9].

As it turns out, not all patients seen in academic medical centers will freely consent to trainee participation in their care, thereby raising a tension between the need to educate the trainee and the requirement to respect patient autonomy. For instance, a recent study reported that 62% of preceptors at one institution "sometimes" or "often" have had patients decline medical student involvement in gynecologic examinations [10]. Given mounting time pressure, such refusals might erode trainee access to sufficient exposure to the breadth and depth of issues in a clinical rotation. Some argue that choosing to use medical services at an academic institution presumes acceptance of the presence of learners; however, only explicit consent with an accompanying right of refusal is adequate to informed consent. Furthermore, for many patients, academic medical centers may be their only source of care, making more problematic any justification for trainee participation based on presumed choice of venue for health care delivery.

Indeed, key to this conclusion is the requirement to obtain informed consent. As noted by the American College of Obstetricians and Gynecologists' Committee on Ethics, obtaining informed consent has an ethical and legal basis, and pertains not just to consent for the provision of medical care or conduct of research, but for the "participation in teaching exercises involving students and residents." It is a process that should result in a patient's understanding (*comprehension*) of her situation and alternatives for treatment, as well as intentional or voluntary choice (*free consent*) to the provision of care or intervention [11]. While it may be appropriate and helpful to discuss with the patient their reasons for refusal of trainee participation, consent for participation should be elicited and informed refusals of trainee participation based on trainee status should be respected.

A second set of issues accentuated by the changing environment of academic medicine raises issues for both clinician educators and *clinician-scientists*. Given the growing role of industrial development of drugs and devices, relationships with industry are being recognized as more common, more important, and more ethically problematic. Sixty percent of research is privately funded, of which 70% is directly supported by the pharmaceutical industry. Fifty percent of CME is funded primarily through educational grants from the pharmaceutical industry [12, 13]. More than one quarter of physicians receive payment for consulting or serving on an advisory board or speakers bureau.

While presently integral to academic medicine – as both a teaching and research endeavor, the involvement of industry has also emerged as cause for concern. The heart of the concern are the different primary loyalties of corporations and physicians – the former, to stockholders and boards; the latter, to patients. The interests of each do not always align. When members of industry interact with clinicians, corporate activities may also generate biases or obligations unrelated to product merit, creating the actuality or appearance of conflict of interest. In fact evidence has accumulated that the provision of even small gifts may shift physician behavior or prescribing practices, not necessarily in the best interests of their patients – or in the interests of the students and residents they may teach [14]. These shifts are characteristic even of situations in which providers do not recognize admit to any changes in practice patterns or prescribing behavior.

Accumulating data about the influence of industry over physician behavior has led several professional and regulatory organizations to develop guidelines about acceptable relationships with industry following recommendations of national leaders [15, 16]. In 2008, the American College of Obstetricians and Gynecologists joined other professional organizations in signaling caution regarding such relationships [17]. In its Committee Opinion, it cautioned against acceptance of gifts even of nominal value; recommended disclosure of financial interests to patients and colleagues; and noted that support from industry for CME and professional meetings should not be directly accepted by physicians; when it is used to support educational activities, disclosure is required. Several academic medical centers have gone further, explicitly prohibiting faculty (both full and, increasingly, adjunct faculty) from participating in certain activities, such as speakers bureaus, in which content is often dictated by companies rather than the faculty that participate. For example, in early 2010, Stanford University extended guidelines to adjunct faculty, based on the idea that anyone who serves medical students in a teaching capacity must present their own work or the medical literature in an unbiased way [18].

Clinician scientists face an array of other issues stemming from the changing landscape of academic medicine, at least two of which deserve mention, and are intimately related. The first are ethical challenges stemming from ever more rapid and widespread innovation and the blurred distinction between innovate practice and formal research such progress often entails; the second are the ethical challenges to the responsible inclusion of pregnant women in research. While not particular to obstetrics and gynecology, the blurring of boundaries between research and innovative practice is particularly common in the specialty due to several factors [19, 20]. First is the clinical context. A significant proportion of innovation in reproductive medicine involves medical *procedures* which lack the adequate regulatory mechanisms for ensuring safety and efficacy that apply to drugs. Development of surgical techniques often occurs in the context of a clinical encounter rather than a research study, and new devices as well as operation-specific "kits" for surgical use are often rapidly and widely adopted before being evaluated by rigorous studies [21]. Second is the political context. Divisive debates about the moral status of embryos and fetuses have led to policies that preclude federal funding of any research in which an embryo is destroyed [22]. As a result, most clinical innovation in reproductive endocrinology has been conducted without the oversight of the federal government, and has instead been funded by private industry or conducted in the context of clinical care. And third is the cultural context, specifically what have been called "protectionist" inclinations and policies toward pregnant women, children and fetuses [23], and a focus on eliminating, rather than responsibly managing, risk in the context of pregnancy [24]. While the reasons for such a stance are beyond the scope of this chapter, it is important to note that the inclination to protect women, fetuses and embryos *from* research rather than *through* research has led to a dearth of studies regarding therapeutics in the populations for whom obstetricians and gynecologists care, and a parallel dearth of data upon which to base treatment decisions [25].

Distinguishing between innovative practice and research is ethically important for at least three reasons. First, again, is the requirement to respect patient autonomy and obtain informed consent. Disclosure of the degree to which the new and novel is also unproven is necessary (although not sufficient) to informed decision making for patients. Second is the need for adequate protection from the harms of novel intervention; in terms of ethics, it reflects the principle of *beneficence*, which holds that the provision of research or care should promote well-being or health [4]. Oversight of research is highly regulated in the US, with research protocols requiring approval by an Institutional Review Board (IRB). Such oversight has been designed in part to advance adequate disclosure of the risks of intervention and ensure through external assessment that risk-benefit calculus is appropriate [26].

The third reason that distinguishing innovative practice from research is ethically important has to do with *justice*. Justice requires that the distribution of the benefits (and burdens) of health interventions be fair, and the provision of care and the conduct of research not be discriminatory. Moreover, newer theories of justice bring to light other problems of the current approach to knowledge and research in reproductive medicine. These theories hold that oppression – in addition to distribution – qualifies as a concern of justice, and highlight women's disadvantage with regard to evidence-based therapeutics as a dimension of women's generally oppressed status in society [27]. Compared with that for other specialties, the burdens of ignorance in obstetrics and gynecology are particularly profound, raising concern that reproducing women and children's interests have not been fairly represented. Because reproductive medicine has often advanced outside of the context of research, there is a dearth of information about evidence-based therapeutics across the breadth of women's reproductive health.

For instance, recent reports have highlighted a profound dearth of information about the impact of assisted reproductive technologies on the health of both children who result from them and women who undertake them [28–31]. A 2008 Cochrane review of outcomes of interventions used in the treatment of infertility indicated that most randomized studies conducted since 2000 either did not measure the short or long term health impact of fertility treatment on women, or did not have sufficient power to detect meaningful differences in delivery rates or obstetric outcomes, particularly less frequent outcomes such as complications affecting maternal health [32].

The status of evidence based therapeutics in established pregnancies is similarly problematic [33]. Because pregnant women are presumed to be excluded from most research studies, evidence about how to safely and effectively treat women's medical illnesses during pregnancy is lacking. This includes a dearth of pharmacokinetic data to guide dosing, effectiveness data to guide medication choice, and information about the safety of medications for pregnant women and fetuses [34]. The result is harm to the health and well being of women and children alike. Many have raised justice as a central concern: current policies have excluded a class of potential beneficiaries (pregnant women) from the immediate and long term benefits of research. Advancing responsible inclusion of pregnant women in research is a moral imperative – and one in which faculty in academic medical centers can and should take a leadership role [35]. Specific strategies, such as increased focus on opportunistic pharmacokinetic studies, ensuring assessment of women's health outcomes in observational studies [36], and advancing legislative strategies akin to those of the pediatrics community (described in more detail elsewhere [33]) are among those that should be advanced by clinician-scientists in academic obstetrics and gynecology.

9.4 Challenges for Institutions: An Ethical Workplace

One of the most profound shifts the landscape of academic obstetrics and gynecology are the demographics of its faculty – perhaps the most striking of which is gender. Less than half a century ago, obstetrics and gynecology was a specialty dominated by men. Not so anymore. As described in Chapter 3, the number of women in the specialty has inclined steadily in recent decades, with females now constituting more than half of full-time faculty at academic institutions. Indeed, the proportion of women who are faculty is higher in obstetrics and gynecology than it is in any other clinical or basic science department [6].

At first glance the shift appears a welcome corrective to the gender imbalance that previously characterized faculty demographics in women's healthcare. But other statistics have been cause for concern. Compared with their male counterparts, women were more likely to be junior faculty, with only 9.1% having reached full professorship compared with 29% of their male counterparts, ratios that have not changed significantly in the last decade. In part attributable to the still relative newness of gender equity in faculty demographics, a contributing factor is also the significant attrition rate of female faculty: over a 10 year time period, only 36% of entry level assistant professors remain at their school, with a vast majority of those leaving the institution having left academia altogether. Attrition is not exclusively a problem of women faculty, although women are especially likely to leave academia when compared to their male counterparts; nor is it exclusively a problem for departments of obstetrics and gynecology, given other departments combined can claim a similarly concerning 10 year retention rate of 44%. But the combination of high attrition among females and a now predominantly (at least at its lower ranks) female specialty, raises ethical challenges for academic obstetrics and gynecology, and call for close attention to the ways that gendered norms may be contributing to the problem.

A central challenge for academic medical centers is developing what might be called an "ethical workplace," which, at the very least, is a workplace where individuals are treated respectfully and fairly with equal opportunities for advancement for faculty regardless of race or gender. This requires not simply a sincere commitment to meritocracy but attention to the sometimes subtle ways that gender shapes the way faculty are assessed, supported or promoted.

Two related conceptions of the role of gendered norms illuminate challenges and opportunities for crafting an ethical workplace. The first is *androcentrism* – which relates to the ways that human society tends to be male-centered. Under androcentrism, man is treated as the tacit standard for what it is to be human – the unstated point of reference. This, theorists have argued, places man in an unfairly privileged position, particularly in what Maggie Little has called the "disturbing cumulative effect on our understanding of "human": over time, our substantive conception of what is normal for humans has come to be filled in by what is normal for men" [5]. Thus man's psychology, body and biography have subconsciously come to be regarded as constituting a human norm. Consider how this might play out in distorting conceptions of the ethical workplace. The academic tenure system provides a vivid example. Many have presumed that evaluating faculty for tenure after 7–10 years should provide a rough indication of productivity or performance over their academic career. But this presupposition assumes a *man's* career. Given 4 years of medical school and four of residency, for women these years correspond to childbearing years (read: pregnancy, birth, breastfeeding, rearing young children) – and are arguably the years *least* likely to represent a woman's future academic potential.

A second challenge to crafting the ethical workplace are *gender schemas*. These refer to unconscious cognitive representations of men as capable of independent and focused action and acting according to reason, and of women as nurturing, communal and acting according to emotion [37]. Notably, stereotyped behaviors about traits and behaviors of good leaders are more agentic and thus more aligned with assumptions about male, rather than female attributes [38]; studies consistently find that men are assumed to posses intrinsic traits that make them better leaders than their female counterparts, even when their credentials are identical. Such schemas have resulted in overrating men and underrating women in professional settings, and have exerted subtle influence across academic medicine, including bias in the selection process for prestigious federal grants [39].

Clearly such biases – often unintended and unconscious – are potential sources of frustration and barriers to advancement, likely to affect the experience of women (and men) in the setting of academic medical centers. While the reasons for faculty members leaving academic obstetrics and gynecology are complex, several studies suggest that the ethical workplace as such has not been broadly achieved – neither in academic medical centers, nor in the American workplace more broadly construed; highlighted here a handful of the larger breadth significant findings.

First are studies that, in exploring reasons behind attrition, identify problematic workplace culture and priorities as a driving force. In contrast to media portrayals of high achieving women's decisions to leave careers as representative of a "sea-change" among daughters of the feminist revolution and a return to (even embracing of) traditionalism, rigorous social science suggests that other issues are at play [40]. For instance, sociologist Pamela Stone, who interviewed a large series of women who chose to leave their careers just as they were on the brink of (or realizing) leadership potential were not in fact "opting out." Rather, she discovered that women "quit because of work, not home – and only as a last resort." Women, she argued, don't "opt out" but are pushed out, and often as they approach the payoff after years of hard work – as they ought just to be taking on the leadership positions they worked so long to assume.

Studies of women specifically in academic medicine have corroborated Stone's findings, indicating that characteristics of the workplace are a significant factor in decisions to leave for non-academic careers or to leave the profession entirely. These include feelings of isolation, lack of role models, lack of formal and informal mentorship, an environment perceived as denigrating to women, frank gender discrimination, and inadequate institutional support for family issues that – despite progress – still fall predominantly on women [41].

Work in psychology has helped to further delineate the sources of frustration. In her book *Necessary Dreams* Anna Fels argues that in order for individuals to pursue their ambitions, it is critical for them to experience *recognition* in their lives, and that the pursuit of mastery requires a particular sort of recognition, which she describes as follows:

> Recognition means being valued by others for qualities that we experience and value in ourselves; it involves appreciation by another person that feels accurate and meaningful to the recipient. Because recognition affirms a person's individual experience or accomplishment, it is different from other forms of attention [42].

Notably, Fels argues that there are gender differences in the ways that recognition is both experienced and dispensed. She points to a large body of evidence that women get less recognition than men – even when their behaviors or achievements are identical. It is possible, then that the continued exodus from academic obstetrics and gynecology can be traced, in part, to gender differences around the dispensation of recognition within academic medical centers.

More concretely, researchers on gender and academic medicine have cited the hierarchical structure of academic medicine, including the indeterminate tenure of department chairs, as having a significant impact on faculty work experience, including advancement, particularly for women [43]. The authors concluded that medical schools should consider alternative models of leadership and managerial styles, including fixed terms for chairs with a greater emphasis on inclusion as a structural reform that could increase opportunities for advancement especially for women in academic medicine.

While far from an exhaustive review of the complex factors that underlie the "leaky pipeline" in academic obstetrics and gynecology, it makes clear that institutions have an important role – beyond ensuring equity of pay and absence of explicit sexual harassment – in retention of male and female faculty alike. This will require more than giving women the opportunities that that their male counterparts have long enjoyed. Rather, since the traditional workplace was shaped by a particular set of needs and priorities (that of men, usually with partners in the home) the traditional workplace may not be an ethical workplace. Crafting the latter will require similar attention to the needs and priorities to the population it now serves – a population in which men and women are both represented.

9.5 Ethical Challenges for the Profession

The changing landscape of academic medicine also presents significant challenges for the profession of obstetrics and gynecology as a whole – challenges which stretch beyond institutions to issues for the professional organization and its leadership.

A central concern for the profession of obstetrics and gynecology as a whole is meeting the increased demand for women's health care services. While the degree to which the health of all women is the profession's responsibility might be debated, a broadly accepted view of medial professionalism holds that the responsibility is strong and fundamental. Because obstetrician-gynecologists often have an exclusive right to offer many health services – a "collective monopoly" – the profession as a whole also has a duty or responsibility to ensure services are readily available to all [44, 45]. In other words, the exclusive rights that come with the practice of obstetrics and gynecology creates a related vulnerability in the population, and generates duties to ensure access to services. This so-called "gatekeeper paradigm" is relevant to at least two ethically controversial issues relating to patient access and facing the specialty of obstetrics and gynecology: approach to conscientious refusals [46], and issues of birth location and attendant. It is likewise relevant to issues of access to care and socioeconomics, discussed in depth elsewhere (Chapter 8).

Consider first the highly controversial topic of conscientious refusals by physicians to provide standard services. Though relevant across fields, the issue has gained particular attention in obstetrics and gynecology in which morally contentious services around reproduction are common [47]. The central question is whether and under what conditions physicians may refuse to provide requested or indicated care if doing so would constitute a significant moral conflict. The debate has been considerably polarized, with one side claiming that physicians who cannot provide such services leave the profession [48]; and the other claiming that physicians have a virtually absolute right to refuse [49].

The latter view was promulgated in the last days of the Bush Administration, with a rule providing sweeping protection for providers who refuse. Specifically, the law protects "actors who have a reasonable connection to the procedure, health service or health service program, or research activity to which they object . . . encompassing individuals who are members of the workforce of the Department-funded entity performing the objectionable procedure," from "participation in any activity with a reasonable connection to the objectionable procedure, including referrals, training, and other arrangements for offending procedures [50]."

The Rule has been sharply and broadly criticized as flying in the face of civil rights law and standards of professionalism [51, 52]. Critics have highlighted the potentially serious consequences of these expansive (and expensive) provisions. Of perhaps most immediate concern is the likelihood that it will further undermine patients' access to necessary health care and information, with excess burden on already disadvantaged populations. Others highlight the extent to which the provisions are inconsistent with professional ethical responsibilities to provide non-discriminatory health care and are likely to cause confusion. Indeed many professional organizations have crafted a compromise position, which holds that providers have a limited right to refuse as a long as access to requested care is maintained, and take a stance that argues for "balance" between the needs of providers and patients [53–55]. These guidelines are all characterized by a trio of requirements that protect patient access: a requirement to provide accurate and complete information, a requirement to provide referral to another provider or health care facility if indicated or requested care constitutes a conflict, and a requirement to provide care regardless of objections in emergency circumstances.

The debate is ongoing, and will likely continue to be a source of contention in the near future. Within weeks of taking the nation's helm, the Obama Administration initiated a proposal to rescind the HHS rule, requested comments in order to evaluate whether to rescind the rule in its entirety or in part [56]. Most proposals for ongoing progress highlight the role of professional organizations in crafting a solution that meets the needs of both professionals and patients. Indeed, as one leading scholar has argued, "accepting a collective obligation does not mean that all members of the profession are forced to violate their own conscience" [46]. ACOG's Ethics Committee argued, rather, that "institutions and professional organizations should work to create and maintain organizational structures that ensure nondiscriminatory access to all professional services and minimize the need for individual practitioners to act in opposition to their conscience" [47]. In February 2011, key portions of the Rule were rescinded [57].

The profession's responsibility in this case stems not just from protecting its individual members' consciences, but from the duty that derives from the profession's monopolistic control over a "public utility" of sorts – reproductive health care. Legal scholar Rebecca Dresser puts it this way: "although every individual member of the medical profession might not have a duty to perform every procedure within his or her competence, society expects the broader profession to adopt reasonable measures to promote patient access to acceptable procedures" [58].

The details of an approach which accomplishes this goal, however, remain a challenge. Legal scholar Fernandez Lynch has proposed a model in which "the government, the profession, and society should allow both doctors and patients to seek out like-minded others" [45]. She pinpoints state licensing boards acting as a proxy for the medical profession as a whole as the most promising organization to render this compromise solution. While beyond the scope of this chapter, her concrete proposal for progress with regard to the ethically fraught issue of conscientious refusal rests squarely on the profession as a whole. Finding a satisfactory solution to the issue of conscientious refusal is both a challenge and responsibility for the profession – clearly a charge for leaders in academic obstetrics and gynecology.

Matching the contentiousness of the debate around conscience are a second set of issues for the profession: issues regarding birth location and provider. Where women should give birth – and by whom they should be cared for in that process, has long been a source of controversy. While the vast majority of births in the US occur in hospitals and are attended by physicians, out of hospital birth has long been advocated by members of the midwifery and women's health advocacy communities as a safe and satisfactory alternative. In several European countries, out of hospital birth is widely accepted; in the Netherlands, 30% of births occur at home. Midwives in European countries play a role in the care of all or most pregnant women, and their involvement is advocated by governmental and professional organizations [59].

The US reflects a different model, favoring the provision of care in hospitals by physicians. While the percentage of births attended by midwives has grown steadily since 1980, it was still only 8% as of 2003. Place of birth is more contentious: both ACOG [60] and the American Medical Association [61] oppose home birth. In an official 2007 policy, ACOG "believes that the hospital, including a birthing center within a hospital complex ... or freestanding birthing centers that meet (certain accreditation) standards is the safest setting for labor, delivery, and the immediate postpartum period" and indicated that it "strongly opposes home birth" and "does not support programs or individuals who provide home births." Several professional and consumer advocacy organization have raised concerns about these policies, citing a lack of evidence to support their position, and the potential to limit women's choices and jeopardize physician back up for women who do choose out of hospital birth. Recently ACOG's Committee on Obstetric Practice issued a Committee Opinion which continued to endorse hospitals and birthing centers as the safest options for women, but emphasized its respect for "the right of a woman to make a medically informed decision about delivery" [62].

The physician shortage accentuates the debate about birth attendants and location – and, notably, further problematizes the aforementioned stance of ACOG and the AMA. By casting a broad net in its attestation about safety according to birthplace, it delineates a broad monopoly over maternity care across the US population. According to the "gatekeeper paradigm" such purported control brings with it responsibility to ensure access to services. Yet access to safe birth – not to mention as access to the type of birth that many childbearing women desire – is currently limited. The reasons are several, and include obstacles to obtaining maternity care, such as poverty, discrimination, and lack of insurance overage, to name a few. In addition, a shortage of obstetrical providers in many regions has contributed to the problem.

While many of these barriers to safe maternity care are not the *fault* of the profession, the profession's claims about appropriately safe maternity care make access to adequate care their *responsibility*. Obviously one approach is to address the physician shortage by training more obstetricians. In addition, however, fostering collaboration with other obstetrical care providers, such as midwives, and pursuing safe (and often preferred) alternatives to hospital birth provide potential solutions that are both patient-centered and cost-effective. Here again, the "gatekeeper model" suggests that given their virtual monopoly on maternity care services, the obstetrical community is in fact responsible for ensuring access to the breadth of services that patients need and want. Doing so may involve a process of intensive collaboration and serious considerations of different models – and locations – for the provision of maternity care.

9.6 Conclusion

Academic medicine in the new millennium presents a breadth of ethical challenges for obstetrics and gynecology – for clinician-scientists and educators; for academic institutions; and for the profession broadly. As is clear from analysis of just a slice of the considerable breadth of ethical issues and challenges for academic obstetrics and gynecology, the privilege of caring for women's across their lifespan brings with it challenges and responsibilities. These include promoting the best interests of patients receiving care in teaching institutions, crafting an ethical workplace for providers of such care, and advancing systems of health care that ensure access to services for all women.

References

1. World Health Organization. 2009. *Women and health: Today's evidence, tomorrow's agenda.* Geneva, Switzerland: World Health Organization. Available at http://whqlibdoc.who.int/publications/2009/9789241563857_eng.pdf.
2. Jonsen, A.R. 1998. *The birth of bioethics*. New York, NY: Oxford University Press.
3. American College of Obstetricians and Gynecologists Committee on Ethics. 2007. Committee Opinion #390: Ethical decision making in obstetrics and gynecology. *Obstetrics & Gynecology* 110(6):1479–87.
4. Beauchamp, T., and J. Childress. 2001. *Principles of biomedical ethics*. New York, NY: Oxford University Press.
5. Little, M. 1996. Why a Feminist approach to bioethics? *Kennedy Institute of Ethics Journal* 6(1):1–18.
6. Rayburn, W.F., B.L. Anderson, J.V. Johnson, M.A. McReynolds, and J. Schulkin. 2010. Trends in the academic workforce in obstetrics and gynecology. *Obstetrics & Gynecology* 115(1):141–146.
7. Accreditation Council for Graduate Medical Education. 2003. *Common program requirements for duty hours*. Chicago, IL: ACGME.
8. Faden, R.R., and T.L. Beauchamp. 1986. *A history and theory of informed consent*. New York, NY: Oxford University Press.
9. American College of Obstetricians and Gynecologists Committee on Ethics. 2007. Committee opinion 385: The limits of conscientious refusal in reproductive medicine. *Obstetrics & Gynecology* 110:1203–1208.
10. Tang, T.S., and E.P. Skye. 2009. When patients decline medical student participation: the preceptor's perspective. *Advances in Health Sciences Education* 14:645–653.
11. American College of Obstetricians and Gynecologists Committee on Ethics. 2009. Committee opinion number 439: Informed consent, August 2009.
12. Studdert, D.M., M.M. Mello, and T.A. Brennan. 2004. Financial conflicts of interest in physicians' relationships with the pharmaceutical industry—self-regulation in the shadow of federal prosecution. *New England Journal of Medicine* 351:1891–1900.
13. Katz, D., A.L. Caplan, and J.F. Merz. 2003. All gifts large and small: Toward an understanding of the ethics of pharmaceutical industry gift-giving. *American Journal of Bioethics* 3(3): 39–46.
14. Wazana, A. 2000. Physicians and the pharmaceutical industry: Is a gift ever just a gift? *Journal of the American Medical Association* 283:373–380.
15. Brennan, T.A., D.J. Rothman, L. Blank, D. Blumenthal, S.C. Chimonas, J.J. Cohen, J. Goldman, J.P. Kassirer, H. Kimball, J. Naughton, and N. Smelser. 2006. Health industry practices that create conflicts of interest: A policy proposal for academic medical centers. *Journal of the American Medical Association* 295(4):429–433
16. Rothman, D.J., and S. Chimonas. 2008. New developments in managing physician-industry relationships. *Journal of the American Medical Association* 300(9):1067–1069.
17. American College of Obstetricians and Gynecologists Committee on Ethics. 2008. Committee opinion #401: Relationships with industry. *Obstetrics & Gynecology* 111(3):799–804.
18. Singer, N. 2010. Stanford Medical School to expand ethics rules. *New York Times*, 21 Mar 2010.
19. Lyerly, A.D. 2009. Review of test tube families: Why the fertility market needs regulation, by Naomi Kahn. *New England Journal of Medicine* 310:429–430.

20. Lyerly, A.D. 2010. Marking the fine line: Ethics and the regulation of innovative technologies in human reproduction. *Minnesota Journal of Law, Science and Technology* 11(2):685–712.
21. Wall, L. 2010. The perils of commercially driven surgical innovation. *American Journal of Obstetrics and Gynecology* 202:30.e1–4.
22. Knowles, L., and G. Kaebnick, eds. 2007. *Reprogenetics: Law, policy and ethical issues*. Baltimore, MD: Johns Hopkins University Press.
23. Mastroianni, A.C., R.R. Faden, and D. Federman, eds. 1994. *Women and health research: Ethical and legal issues of including women in clinical studies*. Washington, DC: National Academy Press.
24. Lyerly, A.D., L.M. Mitchell, E.M. Armstrong, L.H. Harris, R. Kukla, M. Kuppermann, and M. Little. 2009. Risk and the pregnant body. *Hastings Center Report* 39(6):34–42.
25. Lyerly, A.D., M.O. Little, and R. Faden. 2008. Pregnancy and clinical research. *Hastings Center Report* 38(6):53.
26. American College of Obstetricians and Gynecologists Committee on Ethics. 2006. Committee opinion #352: Innovative practice: Ethical guidelines. *Obstetrics & Gynecology* 108(6): 1589–1595.
27. DeBruin, D. 1994. Justice and the inclusion of women in clinical studies: A conceptual framework. In *Women and health research: Ethical and legal issues of including women in clinical studies, vol 2*, eds. A. Mastroianni, R. Faden, and D. Federman D. Workshop and Commissioned Papers, Institute of Medicine. Washington, DC: National Academies Press, pp. 127–150.
28. Rosenwaks, Z., and K. Bedikson. 2007. Further evidence of the safety of assisted reproductive technologies. *Proceedings of the National Academy of Sciences* 104:5709–5710.
29. Manipalviratn, S., A. DeCherney, and J. Segars. 2009. Imprinting disorders and assisted reproductive technology. *Fertility and Sterility* 91:305–315.
30. Reefhuis, J., M.A. Honein, L.A. Schieve, A. Correa, C.A. Hobbs, S.A. Rasmusses, and the National Birth Defects Prevention Study. 2008. Assisted reproductive technology and major structural birth defects in the United States. *Human Reproduction* 24:360–366.
31. Kolata, G. 2009. Picture emerging on genetic risks of IVF. *New York Times*, 17 Feb 2009.
32. Myers, E.R., D.C. McCrory, A.A. Mills, T.M. Price, G.K. Swamy, J. Tantibhedhyangkul, J.M. Wu, and D.B. Matchar. 2008. Effectiveness of assisted reproductive technology. *Evidence Report Technology Assessment* May(167):1–195.
33. Lyerly, A.D., M.O. Little, and R. Faden. 2008. The second wave: Responsible inclusion of pregnant women in clinical research. *International Journal of Feminist Approaches to Bioethics* 1(2):5–22.
34. Chambers, C.D., J.E. Polifka, and J.M. Friedman. 2008. Drug safety in pregnant women and their babies: Ignorance not bliss. *Clinical Pharmacology & Therapeutics* 83(1):181–183.
35. Little, M., A.D. Lyerly, and R. Faden. 2009. Pregnant women and medical research: A moral imperative. *Bioethics Forum* 2:60–65.
36. Lyerly, A.D., M.O. Little, and R. Faden. 2009. The national children's study: A golden opportunity to advance the health of Pregnant women. *American Journal of Public Health* 99:1742–1745.
37. Heilman, M. 2001. Description and prescription: How gender stereotypes prevent women's ascent up the organizational ladder. *Journal of Social Issues* 57:657–674.
38. Valian, V. 2000. *Why so slow? The advancement of women*. Cambridge, MA: The MIT Press.
39. Carnes, M., and C. Bland. 2007. A challenge to academic health centers in the National Institutes of Health to prevent unintended gender bias in the selection of Clinical and Translational Science Award leaders. *Academic Medicine* 82(2):202–206.
40. Stone, P. 2007. *Opting out? Why women really quit careers and head home*. Berkeley, CA: University of California Press.
41. Carnes, M., G. Vandenbosche, P. Agatisa, A. Hirschfeld, A. Dan, J.L.F. Shaver, D. Murasco, and M. McLaughlin. 2001. Using women's health research to develop women leaders in

academic health sciences: The national centers of excellence in women's health. *Journal of Women's Health and Gender-Based Medicine* 10(1):39–46.
42. Fels, A. 2004. *Necessary dreams: Ambition in women's changing lives*. New York, NY: Random House, 9.
43. Conrad, P., P. Carr, S. Knight, M.R. Renfrew, M.B. Dunn, and L. Pololi. 2010. Hierarchy as a barrier to advancement for women in academic medicine. *Journal of Womens Health* 19(4):799–805.
44. Goodin, R.E. 1985. *Protecting the vulnerable: A reanalysis of our social responsibilities*. Chicago: University of Chicago Press.
45. Fernandez Lynch, H. 2008. *Conflicts of conscience in health care: An institutional compromise*. Cambridge, MA: The MIT Press.
46. Charo, A. 2005. The celestial fire of conscience – refusing to deliver medical care. *New England Journal of Medicine* 352:2471–2473.
47. American College of Obstetricians and Gynecologists Committee on Ethics. 2007. Committee Opinion 385: The limits of conscientious refusal in reproductive medicine. *Obstetrics & Gynecology* 110(5):1203–1208.
48. Savulescu, J. 2006. Conscientious objection in medicine. *British Medical Journal* 332(7536):294–297.
49. Curlin, F.A. 2008. *Patient Rights vs. doctor conscience*. Grand Rapids, MI: DeVos Medical Ethics Colloquy.
50. Ensuring That Department of Health and Human Services Funds Do Not Support Coercive or Discriminatory Policies or Practices in Violation of Federal Law, 73 Fed. Reg. 78,072, 78,073 (Dec. 19, 2008) (to be codified at 45 C.F.R. pt. 88).
51. Walker, J. 2009. The Bush administration's midnight provider refusal rule: Upsetting the emerging balance in state pharmacist refusal laws. *Houston Law Review* (46):939–974.
52. National Women's Law Center. 2008. What are some key organizations saying about the proposed HHS rule? Available at http://www.nwlc.org/pdf/keyorgsonhhsrule.pdf.
53. American Medical Association, Council on Ethical and Judicial Affairs. 2007. Report 6-A-07: Physician objection to treatment and individual patient discrimination (Resolution 5, A-06).
54. UK General Medical Council. 2008, Mar. Personal beliefs and medical practice: Guidance for doctors. Available at http://www.gmc-uk.org/guidance/ethical_guidance/personal_beliefs.asp#1.
55. American Academy of Pediatrics Committee on Bioethics. 2009. Policy statement – Physician refusal to provide information or treatment on the basis of claims of conscience. *Pediatrics* 124(6):1689–1693.
56. Stout, D. 2009. Obama set to undo "conscience" rule for health workers. *New York Times*, 27 Feb 2009.
57. National Women's Law Center. 2011. HHS Rescinds Portions of HHS Rule that Threatened Women's Health. Available at http://www.nwlc.org/resource/hhs-rescinds-portions-health-care-denial-rule-threatened-women%E2%80%99s-health#HHS%20PDF.
58. Dresser, R.S. 1994. Freedom of conscience, professional responsibility, and access to abortion. Freedom of conscience, professional responsibility, and access to abortion. *Journal of Law Medicine and Ethics*. 22(3):280–285.
59. Rooks, J. 1997. *Midwifery and childbirth in America*. Philadelphia, PA: Temple University Press.
60. American College of Obstetricians and Gynecologists Executive Board. 2007, May. Statement of Policy: Home births in the United States.
61. American Medical Association House of Delegates. 2008. Resolution A-05, 208. Statement on home deliveries. Available at http://www.ama-assn.org/ama1/pub/upload/mm/471/205.doc
62. American College of Obstetricians and Gynecologists Committee on Obstetric Practice. 2011. Committee opinion #476: Planned home birth. *Obstetrics and Gynecology* 117(2):425–8.

Chapter 10
Preparing for Tomorrow in Academic Women's Health Care

William F. Rayburn and Jay Schulkin

Abstract This summary will serve as a reference in promoting common data definitions across other healthcare fields and in stimulating an adequately funded, better organized national agenda focused on other academic physician workforce issues. The chapter combines conclusions from each chapter to project changes in the academic women's health care landscape as a foundation for continued surveillance and reporting.

Keywords Academic · Workforce · Projections women's health

10.1 Introduction

Numerous pressures at academic health centers are causing women's health care to evolve and redefine itself. Given their long-term commitment and perspective, academic departments have an obligation to lead the process of change while they struggle to adapt to health care reform. A proactive stance will help improve the effectiveness of training programs in women's health care and enhance our influence on patient care and research in women's health.

At the average academic health center today, funding for education and research depends largely on the profitability of practice activities, which represent more than half of the average clinical department's budget [1]. Although many training programs have felt this impact, Medicare indirect graduate medical education payments have helped subsidize these educational responsibilities. As explained in Chapter 8, it has been widely held that free access to patients depends on a primary care designation, except perhaps for obstetrics and gynecology, a procedure-driven specialty in which prevention and coordination of care are emphasized.

W.F. Rayburn (✉)
Department of Obstetrics and Gynecology, University of New Mexico School of Medicine, Albuquerque, NM, USA
e-mail: wrayburn@salud.unm.edu

W.F. Rayburn, J. Schulkin (eds.), *Changing Landscape of Academic Women's Health Care in the United States*, International Library of Ethics, Law, and the New Medicine 48, DOI 10.1007/978-94-007-0931-7_10,

We face significant and rapid changes in preparing for tomorrow's needs. Academic departments have opportunities in women's health care to create a unified curriculum for medical students, share traditional expertise in residency programs, instill concepts of lifelong learning in our graduates, and develop or refine basic and clinical research programs. Presented in this chapter are key findings and implications from prior chapters as the landscape of academic women's health care continues to evolve.

10.2 Trends in Medical School Enrollment and Residency Matching

As noted in Chapter 1, this year marks the 100th anniversary of the Flexner report, *Medical Education in the United States and Canada* [2]. This landmark report, published in 1910, found that medical schools did not base their training on sufficiently high standards of science. While not focusing on women's health care, that report resulted in the closure of many medical schools and caused the number of medical graduates per year to drop precipitously. From 1900 to the mid-1960s, there was almost universal agreement that the nation did not have enough physicians, although the per capita supply remained relatively stable. This perceived shortage was not supported by any studies conducted with acceptable analytic rigor.

The Flexner Report tended to emphasize excellence in isolation. What is required now is the sharing of information with more social interactions amongst physicians. A culture of research needs to continue by sharing resources including knowledge [3–5]. Other lessons learned from Flexnerian sensibilities to apply to today and tomorrow's needs in medical education are reflected in Table 10.1.

Support from federal and state governments, beginning around 1963, resulted in the building of some 40 new medical schools and the expansion of established schools by 1976. By the early 1980s, the number of entry-level medical students plateaued and the construction of new schools ceased due to a projected oversupply of physicians [6]. There remain over 42,000 applicants each year, or nearly 2 1/3 for every available seat in medical school classes. The total enrollment of first-year U.S. medical students (now about 18,000 annually) has increased minimally over the past 25 years [7]. Women now constitute nearly half of all applicants and enrollees (48%).

While medical school enrollment remained steady (approximately 130 students per medical school class), the population of the United States rose from 227 million in 1980 to 304 million in 2008 – an increase of nearly 80 million people in 30 years. Furthermore, demand on physicians' time has and will increase, especially with an aging population and the epidemics of lifestyle-associated illnesses such as obesity, heart disease, and cancer.

When will we know that we have the "right" number of doctors? A standard method for determining this number of medical school graduates is by calculating the physician-to-population ratio. Currently, there are 286 actively-practicing

Table 10.1 Challenges and recommendations identified by the Carnegie Foundation Report which apply to training in women's health care

	The Carnegie Report 2010	
Theme	Challenges	Recommendations
Standardization and individualization	• Medical education is:	• Standardize learning outcomes through assessment of competencies
	o Not outcomes based	• Individualize learning process, allow opportunity to progress within and across levels when competencies are achieved
	o Inflexible	• Offer elective programs to support the development of skills for inquiry and improvement
	o Overly long o Not learner-centered	
Integration	• Poor connections between formal knowledge and experiential learning	• Connect formal knowledge to clinical experience, including early clinical immersion and adequate opportunities for more advanced learners to reflect and study
	• Fragmented understanding of patient experience	• Integrate basic, clinical, and social sciences
	• Poorly understood nonclinical and civic roles of physicians	• Engage learners at all levels with a more comprehensive perspective on patients' experience of illness and care, including more longitudinal connections with patients
	• Inadequate attention to the skills required for effective team care in a complex health care system	• Provide opportunities for learners to experience the broader professional roles physicians • Incorporate interprofessional education and teamwork in the curriculum
Habits of inquiry and improvement	• Focused on mastering today's skills and knowledge without also promoting knowledge-building and an enduring commitment to excellence	• Prepare learners to attain both routine and adaptive forms of expertise
	• Limited and often pro forma engagement in scientific inquiry and improvement exercises	• Engage learners in challenging problems and allow them to participate authentically in inquiry, innovation, and improvement of care
	• Inadequate attention to patient populations, health promotion, and practice-based learning and improvement	• Engage learners in initiatives focused on population health, quality improvement, and patient safety

Table 10.1 (continued)

Theme	The Carnegie Report 2010 — Challenges	Recommendations
	• Insufficient opportunity to participate in the management and improvement of the health care systems within which they learn and work	• Locate clinical education in settings where quality patient care is delivered, not just in university teaching hospitals
Identity formation	• Lack of clarity and focus on professional values	• Provide formal ethics instruction, storytelling, and symbols (honor codes, pledges, and white coat ceremonies)
	• Failure to assess, acknowledge, and advance professional behaviors	• Address the underlying messages expressed in the hidden curriculum and strive to align the espoused and enacted values of the clinical environment
	• Inadequate expectations for progressively higher levels of professional commitments	• Offer feedback, reflective opportunities, and assessment on professionalism, in the context of longitudinal mentoring and advising
	• Erosion of professional values because of pace and commercial nature of health care	• Promote relationships with faculty who simultaneously support learners and hold them to high standards
		• Create collaborative learning environments committed to excellence and continuous improvement

Source: Cooke et al. [15]

physicians per 100,000 people (1 for every 350 persons) in the United States. Based on the rising population, by 2050 we may need 1.2 million doctors in the United States to maintain this ratio. Assuming that the average physician practices for 40 years from medical school graduation until retirement and that American medical schools will not increase their enrollment, our schools will be able to supply slightly more than half of the physicians needed in 2050.

Recognizing this projected shortage of physicians, the Association of American Medical Colleges (AAMC) announced in 2006 the need of a 30% increase in medical school enrollment by 2015 using the 2002 first-year class size as a baseline (see Chapter 2) [8]. Roughly 5,000 new student positions per year will be necessary to graduate 21,000 students to keep up with the nation's growing population and the unacceptably high number of U.S. students training at foreign medical schools. Expansion of the 125 Liaison Council on Medical Education (LCME) accredited medical schools at the time of the recommendation was expected to comprise 76% of the anticipated growth in first-year student enrollments by 2013, while new schools (five accredited as of 2008 and five applicant schools) would account for the remaining 24% of the increase.

After interviewing at selected residency programs, students in their senior year of medical school will enter into the National Residency Matching Program (NRMP), and eventually match with training institutions. In recent years, approximately 5.0% of all U.S. medical students pursued residencies in ob-gyn programs, 15.1% into family and community medicine, and 5.8% into internal medicine [9]. First-year resident positions will be filled by graduates who trained outside U.S. allopathic medical schools (international medical graduates, osteopathy medical schools) in 26% residency slots in ob-gyn, 56% slots in family and community medicine, and 51% slots in internal medicine. At the same time, hospitals and other health care institutions will fill more than one quarter of the nation's 100,000 available residency positions with international medical graduates (IMGs) [8]. At the same time, first-year enrollment at U.S. osteopathic schools in 2002 (3,079 students) increased in 2008 by 53.7% (to 4,732 students) and is projected to increase in 2013 by 79.2% (to 5,519 students).

10.3 Need for Additional Graduates in Obstetrics and Gynecology

Despite the rising population in the United States, the number of US medical school graduates entering many residency programs especially in obstetrics and gynecology has not increased. The number of first year residents has finally equaled the previous peak of 1,218 (7.9% of US medical school graduates) in 1993. The number of ob-gyn residents has increased minimally (2.4%) since academic year 2001–2002. Compared with training programs in other specialties, residents in internal medicine, family and community medicine, and ob-gyn are widely divergent. For example, obstetrics and gynecology has the highest percent of residents who are female (78 vs. 45%) and the highest percentage of residents who are African American, American Indian/Alaska Native, or Native Hawaiian/Pacific Islanders (11 vs. 6%) [9]. The percentage of residents of Hispanic origin is similar to the other residency specialties (8 vs. 8%).

Understanding tomorrow's needs about an adequate number of well-trained graduates from residency programs requires a brief history of the major role played by government and an awareness of signal events during the past 45 years. Anecdotal evidence antedating the 1960s led to a general consensus that the United States had a shortage of physicians, prompting federal and state governments to allocate funds to increase supply. This support began in the early 1960s, and by 1976 resulted in the building of 40 new medical schools and the expansion of established schools. A signal event in 1965 was the passage of Medicare and Medicaid, when the federal government began to assume the responsibility of providing health care for the elderly and disabled. By funding hospitals to care for the Medicare population, funding was appropriated for the education of resident physicians at those hospitals, triggering a gradual but significant increase in the supply of doctors. Various commissions and studies in the 1980s pointed to an oversupply of physicians, with even larger surpluses projected by 2000 and 2020 [2].

Reforms enacted by the Balanced Budget Act (BBA) of 1997 were intended to curb Medicare expenditures rather than to base Medicare GME policy on workforce requirements [10]. The BBA included changes that altered GME funding which reduced the growth and number of intern/resident training positions while attempting to maintain primary care positions. Instead of paid managed care companies for the care of Medicare enrollees, Medicare paid teaching hospitals directly. Hospitals were encouraged to reduce their residency training programs by 20–25% in general except for primary care residents. To encourage primary care programs, the BBA also expanded funding for facilities (especially off-site) to include rural health clinics and community health centers and Medicare managed care plans.

Probably the single greatest impact on training residents was the limitation on work hours. A typical intern's weekly schedule in the 1960s was more than 100 h. The event that started legislative action in New York State was the death of a patient named Libby Zion in 1985, when resident physician fatigue was suspected to be a contributor. The resultant "80 h work week" applied to all time spent by every resident in a hospital, including the clinic, emergency department, operating room, labor and delivery, and teaching conferences. A resident could be on-call no more than every third day, on average, and must have at least 24 h off each week. These restrictions are expensive and difficult to enforce, violations are not always reported, and concern exists in procedure-based medical fields such as obstetrics and gynecology about time could be insufficient for some in surgical training.

The impact of this new "80 h work week" on reducing patient mortality and morbidity remains unclear. In terms of its impact on the workforce, the fewer "man" h roughly translate to a 20% reduction in work hours of the nation's approximately 112,000 medical residents and fellows, which is equivalent to losing the workload of about 15,000–22,000 full-time positions. As is the case with the "80 h" paradigm, we do not have solid evidence of any deleterious or helpful effects of shortened training programs on patient safety or quality care.

A formidable task lies before us. In the judgment of some national groups studying physician workforce issues, the most severe shortage of physicians could be in obstetrics and gynecology. The Balanced Budget Act should be revised. As with the increase in medical school enrollment, we must increase the number of ob-gyn resident positions to serve a population that is predicted to nearly grow another 40% by 2050. Otherwise, the only expansion would be in the proportion of residency positions filled by US medical graduates rather than by international graduates. This population-based increase in demand will be accompanied by a growing database of knowledge, demand for specialization, more cutting edge technology such as laparoscopic, robotic, and minimally invasive procedures, and a desire of many graduates to specialize so as to stand out in their own niches within our broader health discipline.

The cost of training additional clinicians and surgeons needed in all specialties is slightly less than 500 million dollars per year in current dollars [11]. The greatest costs are thought to be in ob-gyn, orthopedics, urology, and general surgery. While this cost to add new physicians is significant, it is a very small fraction of current Medicare expenditures. It is entirely relevant to consider "what if" scenarios related to the cost of funding different numbers of residency positions depending on the severity of the impending shortage. Despite concerns about escalating costs, this expense in training additional residents should be viewed as an essential investment if women are to have timely access to a physician's care.

10.4 Trends in Academic Departments

Current data and projections about department sizes are important for faculty, the specialty as a whole, and those either considering or beginning an academic career in women's health care. Periodic national reviews of department sizes are essential to determine whether a sufficient faculty workforce is available to fulfill missions in teaching, patient care, and research.

The number of full-time ob-gyn faculty at United States medical schools has more than doubled in the past 31 years [1]. Differences exist between the types of schools. The modest growth in ob-gyn department sizes between 1994 and 2008 was only among physician faculty. In general, private medical schools have more faculty than public institutions, and this difference is accounted for by an increasing number of physicians. The top 40 research-intensive schools, according to 2009 National Institutes of Health funding, had more ob-gyn faculty than at the less research-intensive and at the more community-based medical school departments (see Chapter 5).

The most substantial change in faculty demographics was the increase in number and proportion of faculty who are females (see Chapter 3). For example, the proportion of women per ob-gyn department rose from 14.3% in 1983 to 34.1% in 1994 to now 52.3% – higher than any other medical or surgical specialty including pediatrics. With this gradual but more recent increase, women are more likely than men to be instructors or assistant professors than to be associate professors or professors.

The development of faculty requires financial, personnel, and facility resources to meet department missions. Greater demands on faculty in core clinical departments to teach and provide women's health care make protected time to support grant writing and scholarly activity a constant challenge. As departments expand to meet increasing patient care and educational demands, the clinician educator will play a more central role (see Chapter 4). Long-term retention rates of entry-level faculty at their original departments has been similar for faculty in clinical departments. Among those who leave their department, generalist faculty, especially those who are clinician educators, were much more inclined than specialists to leave academia than to switch schools.

The average percentage of positions filled annually by first-time clinical chairs between 1979 and 2007 is about the same (7–9%) [12]. Regardless of the department, clinical chairs find themselves faced with more financial pressures, information overload, and effort invested in managing and implementing organizational change. Compared with other core clinical departments, retention of first-time ob-gyn chairs declined most consistently over the past 30 years from being the highest to now among the lowest. The median tenure of first-time ob-gyn chairs (7.3 years) is now comparable to chairs in surgery, slightly longer than for internal medicine, and shorter for psychiatry, family and community medicine, and pediatrics [12].

A report by the Association of American Medical Colleges (AAMC) found that between 1988 and 1998, faculty lost or gained the most ground in financial compensation depending on their specialty [13]. Data from the annual AAMC Faculty Salary Survey for academic years 2000–2001 through 2008–2009 revealed that unadjusted compensation improved for ob-gyn faculty by 24.8% (or 3.3% annually) during that 9 year period [14]. Inflation-adjusted growth of faculty salaries was similar for those in general internal medicine, family and community medicine, ob-gyn, general pediatrics, and general surgery.

A limitation of AAMC Faculty Roster is that only full-time faculty are counted. All departments now have more part-time faculty. A national survey of chairs disclosed that this pattern is anticipated to increase [1]. The change in workforce to accommodate more part-time faculty is critical for chairs and deans to track so that faculty needs can be more accurately projected (see Chapter 6). More medical schools are now developing policies to optimize recruitment and promotion of part-time faculty as the changing landscape of academic medicine becomes more dependent on maintaining satisfied faculty members.

10.5 Preparing for Tomorrow's Educational Needs

Preparing for tomorrow's needs in academic women's health care requires change and redesign of educational programs. The Carnegie Foundation issued a report, *Educating Physicians: A Call for Reform of Medical School and Residency* that calls for standardizing learning outcomes and individualizing the learning process, promoting multiple forms of integration, incorporating habits of inquiry and improvement, and focusing on the progressive formation of the physician's professional identity (Table 10.1). Cooke and colleagues wrote this report which traces the seeds of these themes in Flexner's work and describes their own conceptions of them, addressing the prior and current challenges to medical education as well as recommendations for achieving excellence [15]. Their key findings apply to advancement of teaching about women's health care.

Findings presented in the text include an expanding medical school enrollment and workforce statistics that indicate the growing need for more training of generalist physicians interested in women's health. Appropriate training of medical students and residents as eventual providers of coordinated care for women will require an interdisciplinary curriculum; expansion of the number of residency positions; promotion of gender, racial, and ethnic diversity at all levels of medical education and academic leadership; and development of expanded opportunities for research in women's health, supported especially by the National Institutes of Health. The numbers of faculty in family and community medicine, internal medicine, and ob-gyn will inevitably expand to meet these increased demands for undergraduate and graduate medical education as long as funding from a variety of resources is maintained or continues to grow.

Many chapters in this book have highlighted the change in medicine and women's health. From women joining the field of medicine in extraordinary amounts in the U.S., to a culture of medicine nurtured within women's health issues, to an inherent uncertainly of how to retain faculty must ensure that faculty are satisfied and not demoralized by the diverse and conflicting demands. Required in this changing landscape are flexible options in which physicians can participate in the field of medicine in diverse ways: as part-time physicians, hospitalists, and use of an older workforce.

As we look ahead, certain recommendations are to be offered to address those needs in academic women's health care, as set forth by the Advisory Committee on Workforce Studies and Planning, American College of Obstetricians and Gynecologists [16]:

- Encourage more medical students to pursue careers in women's health.
- Promote federal funding of selective increases in residency slots, especially at programs in states where there exists an underrepresentation of women's health care physicians or where there is anticipated to be continued population growth.
- Publicize loan repayment programs and other efforts to attract family physicians, general internists, and ob-gyns to health provider shortage areas.
- Promote sustainable physician-led collaborative models with certified nurse midwives, nurse practitioners, and physician assistants, to establish physician-led "teams".
- Alter practices by improving efficiency, reconfiguring the way certain services are delivered, and increasing the use of health information technology.
- Recognize and respond to faculty life-style concerns (i.e. more flexible scheduling, part-time work).
- Improve data collection about the supply and demand for women's health care physicians.

Medicine is one of the great wonders of human invention. The chapters of this book have noted the reality of a changing profession, and suggested some possibilities for a future of medicine that promotes flexibility, faculty satisfaction, and a culture of social medicine, embodied in the free exchange of resources and information to advance women's health care.

References

1. Rayburn, W.F., B. Anderson, J. Johnson, M. McReynolds, and J. Schulkin. 2010. Trends in the academic workforce of obstetrics and gynecology. *Obstetrics & Gynecology* 115:141–146.

2. Flexner, A. 1910. *Medical education in the United States: A report to the Carnegie foundation for the advancement of teaching No. 4*. Carnegie Foundation for Higher Education, Boston, MA.
3. Irby, D., M. Cooke, and B. O'Brien. 2010. Calls for reform of medical education by the Carnegie foundation for the advancement of teaching: 1910 and 2010. *Academic Medicine* 85:220–227.
4. Kirch, D. 2010. Commentary: The Flexnerian legacy in the 21st century. *Academic Medicine* 85:190–192.
5. Schulkin, J. 2000. Decision sciences and evidenced-based medicine: Two intellectual movements to support clinical decision making. *Academic Medicine* 75:816–818.
6. Summary Report of the Graduate Medical Education National Advisory Committee to the Secretary, Department of Health and Human Services, vols I and III GMENAC Summary Report. 30 Sept 1980.
7. Association of American Medical Colleges Center for Workforce Studies. 2010. AAMC medical school enrollment plans through 2013: Analysis of the 2008 AAMC Survey. Available at: http://www.aamc.org/data/aib/. Accessed 22 Feb 2010.
8. Schulman, S., and E. Salsberg. 2009. Projections of future medical school enrollment. *AAMC Analysis in Brief* 9(3). http://www.aamc.org/data/aib/achissues/aibvol9no3.pdf. Accessed 9 Mar 2010.
9. Brotherton, S.E., and S.I. Etzel. 2010. Graduate medical education, 2009–2010. *Journal of the American Medical Association* 304:1255–1270.
10. Salsberg, E., P. Rockey, K. Rivers, S. Brotherton, and G. Jackson. 2008. U.S. residency training before and after the 1997 Balanced Budget Act. *Journal of the American Medical Association* 300:1174–1180.
11. Williams, T.E., B. Satiani, and E.C. Ellison. 2009. *The coming shortage of surgeons. Why they are disappearing and what that means for our health Santa Barbara, CA*. Santa Barbara, CA: ABC-CLIO.
12. Rayburn, W.F., J.D. Waldman, R. Schrader, A. Fullilove, and J. Lang. 2009. Retention of chairs in obstetrics and gynecology: A comparison with other clinical departments. *Obstetrics & Gynecology* 114:130–135.
13. Studer-Ellis, E., J.S. Gold, and R.F. Jones. 2000. Trends in U.S. medical school faculty salaries 1988–1989 to 1998–1999. *Journal of the American Medical Association* 284:1130–1135.
14. AAMC Faculty Salary Survey. 2009. https://services.aamc.org/fssreports/general/genratable03.cfm?current_survey_year=2009&26taqbel_num=12
15. Cooke, M., D.M. Irby, and B.C. O'Brien. 2010. *Educating physicians: A call for reform of medical school and residency*. San Francisco, CA: Jossey-Bass-Carnegie Foundation for the Advancement of Teaching.
16. ACOG Workforce Taskforce Worksheet. 2011. *The obstetrician-gynecologist*. Washington, DC: ACOG.

Index

W.F. Rayburn, J. Schulkin (eds.), *Changing Landscape of Academic Women's Health Care in the United States*, International Library of Ethics, Law, and the New Medicine 48, DOI 10.1007/978-94-007-0931-7, © Springer Science+Business Media B.V. 2011

FSC
www.fsc.org
MIX
Papier | Fördert gute Waldnutzung
FSC® C083411

Kolibri 360 GmbH
Ferdinand-Jühlke-Straße 7
99095 Erfurt, Deutschland
produktsicherheit@kolibri360.de